Ordinary and Partial
Differential Equations

For

Mathematicians, Physicists, and Engineers

Third Year College Course

By

Mohamed F. El-Hewie

2013

TABLE OF CONTENTS

List of Examples

9

11

INTRODUCTION

This book comprises a course in differential equations, which students of engineering, physics, and mathematics complete as a requirement of **bachelor in science degree**. **The reader must possess basic skills in calculus,** since all elementary differentiations and integrations in this book assume that the student could visually spot the derivation from previous years in high school or college.

The author reproduced his own study notes and comments on methods of solutions in order to offer the reader with the feel and touch of another fellow student thriving to master the philosophy of **mathematical formulation** of physical problems beyond the distilled academic presentation.

In the maze of **differential equations**, the student faces the challenges of sorting out orders of derivatives, degrees of power, homogeneity of variables, variability of coefficients, partiality of derivatives, completeness and exactness of solutions along the road of making mathematical sense. Even though Today's online math solvers offer great help in expedient fashion, learning the methods of integrating differential equations empowers the learner with the riches of calculus in interpreting the physical world. For example, solving a nonhomogeneous differential equation with paper and pencil should give the student greater understanding of the meaning of initiating actions rather than interpreting ongoing actions without learning about their genesis. In addition, solving differential problems of striking similarities with pencil and paper should alert the student to the parallel roles of material entities in the spacetime continuum, such as the flow of heat, electricity, fluids, and gases, all performing differential operations imposed by the laws of nature.

Since I wrote this book after completing a lengthy and daunting book on the Differential Equations of Linear Elasticity of Homogeneous Media, I appreciate the diversity and cursory nature of learning the techniques of solving differential equations rather than meddling with the complicated properties of materials. Nevertheless, spending over eight months studying the theory of elasticity alone gave me the deepest pleasure of sensing the power and **limitation of differential equations**.

The book is organized in the logical fashion as presented to college students. The ordinary differential equations (**o.d.e.**) are first studied in great details, since partial differential equations (**p.d.e.**) must be rendered ordinary by separation of variables so as to yield meaningful solution. When separation of variables is untenable (such as in nonlinear partial differential equations), referrals to numerical solutions are given. Within the scope of o.d.e., first- and second-order differential equations are discussed in details, also since equations of higher orders could be reduced in order by successive methods of substitutions, discussed in the book. Also, within the scope of o.d.e., equations with constant coefficients are dealt with greater details, since variable coefficients could be rendered constants by interim substitutions and reverse substations. Also, dealt with is the reduction of higher degrees of variables to lesser degrees.

In an undergraduate course of such broad nature and slight depth, where equations are solved for the sake of **learning the technique of integration**, the reader should be alerted to the two major troubles with differential equations, which are exactness and completeness. Thus, obtaining a solution of differential equations that satisfy given boundary conditions neither amounts to exactness or completeness of solution unless one could prove that alternate and different solutions cannot exist or that the obtained solutions are the only ones available to a given physical phenomena. A relevant example is the flow of heat in a medium where we always assume that the medium thermal constants, such as diffusivity and specific heat, are independent of the sought dependent function or that the heated particles do not interchange energies among themselves beyond conventional laws of mechanics. In such cases, boundary and initial conditions alone give approximate solutions.

Finally, as this book is intended for undergraduate students, most studied examples are either put in one-dimensional geometry, dealing with homogeneous medium in order to avoid complex nonlinear terms, or dealing with isotropic medium where direction has no effect on physical constants of the medium.

Like most undergraduate courses, each example and each problem has a life of its own, teaching the learner new skills and reducing the burden of comprehensive reading which most math-inclined students dread doing.

Mohamed F. El-Hewie
New Jersey, USA
August 20, 2013
.

CHAPTER 1

FIRST-ORDER

ORDINARY DIFFERENTIAL EQUATION

1-1. Definition of first-order o.d.e.

1. The **order of o.d.e.** depends on the order of the differential derivative of dependent variable, y, with respect to independent variable, x, as follows:		2. The **degree of the first-order o.d.e.** is the power exponent of the first derivative:	
First-order	$\dfrac{dy}{dx}$	First degree	$\dfrac{dy}{dx}$
Second-order	$\dfrac{d^2 y}{dx^2}$	Second degree	$\left(\dfrac{dy}{dx}\right)^2$
Nth-order	$\dfrac{d^n y}{dx^n}$	nth degree	$\left(\dfrac{dy}{dx}\right)^n$

3. The **linearity of o.d.e.** is determined by the absence of the dependent variable; y, from the coefficients of its derivatives with respect to the independent variable; x.

e.g.,   Linear o.d.e.: $\dfrac{d^2 y}{dx^2} + 2xy = 0$

e.g.,   Non-Linear o.d.e.: $y \dfrac{d^2 y}{dx^2} + 2xy = 0$

4. The **homogeneity** of o.d.e. is defined by the absence of terms including the independent variable alone:

e.g.,   Homogeneous o.d.e.: $\dfrac{d^2 y}{dx^2} + 2xy = 0$

e.g.,   Non- Homogeneous o.d.e.: $\dfrac{d^2 y}{dx^2} + 2xy + 2x = 0$

5. First-order o.d.e. may have any of the following **forms:**

(i) $F\left(x, y, \dfrac{dy}{dx}\right) = 0$ (1-1.1)	e.g., $\left(y + xy^2\right)dx - xdy = 0$ (1-1.2)
(ii) $\dfrac{dy}{dx} = f(x, y)$ (1-1.3)	e.g., $\dfrac{dy}{dx} = \dfrac{y}{x} + \tan\left(\dfrac{y}{x}\right)$ (1-1.4)
(iii) $F(x, y)dx + G(x, y)dy = 0$ (1-1.5)	e.g., $x\left(1 + y^2\right)dx - y\left(1 + x^2\right)dy = 0$ (1-1.6)

1-2. Separable first-order o.d.e.

Independent variables and dependent variables can be lumped separately on the opposite sides of the equal sign.

Example 1

Solve the differential equation

$$xdx + ydy = 0 \tag{1-2.1}$$

Solution

Apply the integration operator on both members of (1-2.1), each with respect to its respective integrand, we get

$$\int xdx + \int ydy = C$$
$$\frac{x^2}{2} + \frac{y^2}{2} = C \tag{1-2.2}$$
$$y = \sqrt{C'-x^2}$$

C' is a constant. Therefore, separable first order o.d.e. (1-2.1) yields the **equation of a circle** (1-2.2), upon integration.

Example 2

Solve the differential equation

$$x\left(1+y^2\right)dx - y\left(1+x^2\right)dy = 0 \tag{1-3.1}$$

Solution

First, we take advantage of the separatability of variables, lump them in such manner that facilitates separable integrands as follows

$$\frac{x}{\left(1+x^2\right)}dx - \frac{y}{\left(1+y^2\right)}dy = 0 \tag{1-3.2}$$

Then, substitute by $2xdx = dx^2$ and $2ydy = dy^2$ to get

$$\frac{dx^2}{2\left(1+x^2\right)} - \frac{dy^2}{2\left(1+y^2\right)} = 0 \tag{1-3.3}$$

15

Integrating, we get

$$\int \frac{dx^2}{(1+x^2)} - \int \frac{dy^2}{(1+y^2)} = C$$

$$\int \frac{d(1+x^2)}{(1+x^2)} - \int \frac{d(1+y^2)}{(1+y^2)} = C$$

$$\ln(1+x^2) - \ln(1+y^2) = C \qquad (1\text{-}3.4)$$

$$\ln \frac{(1+x^2)}{(1+y^2)} = C$$

$$(1+x^2) = C(1+y^2)$$

$$y^2 + \frac{x^2}{C} = C' \qquad (1\text{-}3.5)$$

Thus, the differential equation (1-3.1) yields the **equation of ellipse**, or circle, depending on the value of the constant C'.

We also note that first order o.d.e. contains only **one constant**.

1-3. Homogeneous first-order o.d.e.

A homogeneous equation differs from equation (1-2.1) in that the differentials are divided by the variables of their kinds, instead of being multiplied.

Example 3

Solve the homogeneous first order o.d.e.

$$\frac{dy}{dx} = f\left(\frac{y}{x}\right) \qquad (1\text{-}4.1)$$

Solution

This could be solved by the following substitution

$$s = \frac{y}{x} \qquad (1\text{-}4.2)$$

Which upon differentiation gives?

$$\frac{dy}{dx} = x\frac{ds}{dx} + s \qquad (1\text{-}4.3)$$

16

Hence, from equations (1-4.1) and (1-.4.3), we get

$$\frac{dy}{dx} = x\frac{ds}{dx} + s = f(s) \tag{1-4.4}$$

As we see, we have rendered equation (1-4.1) separable, with the substitution of (1-4.2), as follows

$$\frac{ds}{f(s) - s} = \frac{dx}{x} \tag{1-4.5}$$

Integrating, we get

$$\int \frac{ds}{f(s) - s} = \int \frac{dx}{x} = C \ln x \tag{1-4.6}$$

Which could be written as

$$x = C'e^{\int \frac{ds}{f(s)-s}} \tag{1-4.7}$$

C' is a constant. Thus, the homogeneous first order o.d.e., (1-4.1), yields upon integration the transcendental function (1-4.7) in which the two variables x and y cannot be represented algebraically in separable form.

Example 4

Solve the differential equation

$$\frac{dy}{dx} = \frac{y}{x} + \tan\left(\frac{y}{x}\right) \tag{1-5.1}$$

Solution

Using the substitutions (1-4.2) and (1-4.4), equation (1-5.1) is expressed in terms of s as follows

$$x\frac{ds}{dx} + s = s + \tan(s) \tag{1-5.2}$$

The cancellation of the free s from both sides gives the separable equations

$$\frac{ds}{\tan(s)} = \frac{dx}{x} \tag{1-5.3}$$

17

This, could be manipulated and integrated as follows

$$\frac{\cos(s)ds}{\sin(s)} = \frac{dx}{x}$$

$$\frac{d\sin(s)}{\sin(s)} = \frac{dx}{x}$$

$$\ln(\sin(s)) = \ln x + C$$

$$\ln\left(\frac{\sin(s)}{x}\right) = C \qquad\qquad (1\text{-}5.4)$$

$$\sin(s) = xC'$$

$$\sin\left(\frac{y}{x}\right) = xC'$$

$$y = x\sin^{-1}(xC')$$

C' is a constant.

Example 5

Solve the differential equation

$$\frac{dy}{dx} = \frac{xy}{x^2 - y^2} \qquad\qquad (1\text{-}5.5)$$

Solution

First, equation (1-5.5) is homogenized by dividing RHS numerator and denominator by x^2, then substituting by equation (1-4.2), we get

$$\frac{dy}{dx} = \frac{xy}{x^2 - y^2}$$

$$= \frac{\dfrac{y}{x}}{1 - \left(\dfrac{y}{x}\right)^2} = \frac{s}{1 - s^2} \qquad\qquad (1\text{-}5.6)$$

Then, substituting from (1-4.3) in (1-5.6), we get

18

$$x\frac{ds}{dx} + s = \frac{s}{1-s^2}$$

$$x\frac{ds}{dx} = \frac{s}{1-s^2} - s \qquad (1\text{-}5.7)$$

$$\frac{\left(1-s^2\right)ds}{s^3} = \frac{dx}{x}$$

Thus, variables in equation (1-5.7) are separated and integration gives

$$\int \frac{ds}{s^3} - \int \frac{ds}{s} = \int \frac{dx}{x}$$

$$-\frac{1}{2s^2} - \ln s = \ln x + C \qquad (1\text{-}5.8)$$

Upon farther simplification, we get

$$\ln xs = C - \frac{1}{2s^2}$$

$$y = C'e^{-\frac{x^2}{2y^2}} \qquad (1\text{-}5.9)$$

C' is a constant.

1-3.1. Homogenization of first-order o.d.e. by translation of origin of coordinates and rotation of axes

Example 6

Solve the differential equation

$$\frac{dy}{dx} = f\left(\frac{a_1x + b_1y + c_1}{a_2x + b_2y + c_2}\right) \qquad (1\text{-}6.1)$$

Solution

Case 1: The lines ($L_1 = a_1x + b_1y + c_1$) and ($L_2 = a_2x + b_2y + c_2$) are not parallel

In order to homogenize equation (1-6.1), we need to eliminate the constants c_1 and c_2 by translating the origin of coordinates to the point of intersection of the two lines.

Substitute by

$X = x - x_0$
$Y = y - y_0$ (1-6.2)

The two new constants x_0 and y_0 are chosen to eliminate c_1 and c_2.

Therefore,

$$\frac{dy}{dx} = \frac{dY}{dX} \qquad (1\text{-}6.3)$$

$$\frac{dy}{dx} = f\left(\frac{a_1 X + b_1 Y}{a_2 X + b_2 Y}\right)$$

$$= f\left(\frac{a_1 + b_1 s}{a_2 + b_2 s}\right) \qquad (1\text{-}6.4)$$

Where,

$$s = \frac{Y}{X} \qquad (1\text{-}6.5)$$

Similar to equation (1-4.3), the differentiation of (1-6.5) gives

$$\frac{dY}{dX} = X\frac{ds}{dX} + s \qquad (1\text{-}6.6)$$

Substituting from (1-6.6) into equation (1-6.4), and renaming the function f to g, with the new arguments, we get

$$X\frac{ds}{dX} + s = g(s)$$

$$X\frac{ds}{dX} = g(s) - s \qquad (1\text{-}6.7)$$

i.e.,

$$\frac{ds}{g(s) - s} = \frac{dX}{X} \qquad (1\text{-}6.8)$$

Which is homogeneous first-degree o.d.e.

Case 2: The lines ($L_1 = a_1 x + b_1 y + c_1$) and ($L_2 = a_2 x + b_2 y + c_2$) are parallel

The two parallel lines L_1 and L_2 has the same slope given by

$$\frac{a_1}{b_1} = \frac{a_2}{b_2} = C \qquad (1\text{-}7.1)$$

Where C is the constant slope of the two lines.

20

Substituting from (1-7.1) into (1-6.1), we get

$$\frac{dy}{dx} = f\left(\frac{Cb_1 x + b_1 y + c_1}{Cb_2 x + b_2 y + c_2}\right)$$

$$= f\left(\left(\frac{b_1}{b_2}\right)\frac{Cx + y + d_1}{Cx + y + d_2}\right)$$

(1-7.2)

Substituting by the new variable

$$s = Cx + y$$
$$\frac{dy}{dx} = \frac{ds}{dx} - C$$

(1-7.3)

Farther, substituting from (1-7.3) into (1-7.2), we get

$$\frac{ds}{dx} - C = f\left(\left(\frac{b_1}{b_2}\right)\frac{s + d_1}{s + d_2}\right)$$

$$= g(s)$$

(1-7.4)

Where the new function g accounts for the altered structure of the argument of the function f.

Therefore, equation (1-7.4), we get

$$\frac{ds}{g(s) + C} = dx$$

(1-7.5)

Thus, the case of parallel lines, in the (1-6.1) yields the **separable variables**, equation (1-7.5).

Example 7

Solve the differential equation

$$\frac{dy}{dx} = \frac{x - y + 1}{x + y - 3}$$

(1-8.1)

Solution

The two lines ($L_1 = x - y + 1$) and ($L_2 = x + y - 3$) intersect at a point determine by solving the two equations

21

$$x - y + 1 = 0$$
$$x + y - 3 = 0 \qquad (1\text{-}8.2)$$

Adding the two equations by member, we get $x_0 = 1$.
Subtracting the two equations by members, we get $y_0 = 2$.

Thus, equation (1-8.1) can be homogenized by substituting from equation (1-6.2) by

$$X = x - 1$$
$$Y = y - 2 \qquad (1\text{-}8.3)$$

Thus, equation (1-8.1) becomes

$$\frac{dy}{dx} = \frac{(X+1)-(Y+2)+1}{(X+1)+(Y+2)-3}$$
$$= \frac{X-Y}{X+Y}$$
$$= \frac{1-s}{1+s} \qquad (1\text{-}8.2)$$

Substituting by the derivative of Y, from (1-6.6), we get

$$X\frac{ds}{dX} + s = \frac{1-s}{1+s} \qquad (1\text{-}8.3)$$

i.e.,

$$X\frac{ds}{dX} = \frac{1-2s-s^2}{1+s}$$
$$\frac{(1+s)ds}{1-2s-s^2} = -\frac{dX}{X} \qquad (1\text{-}8.4)$$

Equation (1-8.4) can be farther simplified to an integrable form as follows

$$\frac{(1+s)ds}{2-(1+s)^2} = -\frac{dX}{X}$$
$$\left(\frac{-1}{2}\right)\frac{d(2-(1+s)^2)}{2-(1+s)^2} = -\frac{dX}{X} \qquad (1\text{-}8.5)$$

Integration gives

$$\left(\frac{-1}{2}\right)\ln\left(2-(1+s)^2\right)=-\ln X+C$$

$$\ln\left(2-(1+s)^2\right)=2\ln X+C'$$ (1-8.6)

$$\ln\left(2-(1+s)^2\right)=\ln X^2+C'$$

$$\left(2-(1+s)^2\right)=C''X^2$$

Finally, restoring the original form in terms of x and y, by substituting from (1-8.3), and by s = Y/X, we get

$$y=2+(x-1)\left(-1\pm\sqrt{2-C''(x-1)^2}\right)$$ (1-8.7)

C" is a constant.

Example 8

Solve the differential equation

$$\frac{dy}{dx}=\frac{2x+y-1}{4x+2y+5}$$ (1-9.1)

Solution

The two lines $(L_1 = 2x + y - 1)$ and $(L_2 = 4x + 2y + 5)$ are parallel. Therefore, substituting by

$$s = 2x + y$$ (1-9.2)

Differentiating, we get

$$\frac{ds}{dx}=2+\frac{dy}{dx}$$ (1-9.3)

Substituting by the derivative of y from (1-9.3), and s from (1-9.2) in (1-9.1), we get

$$\frac{ds}{dx}-2=\frac{s-1}{2s+5}$$ (1-9.4)

i.e.,

$$\frac{ds}{dx}=\frac{5s+9}{2s+5}$$

$$\frac{(2s+5)ds}{(5s+9)}=dx$$ (1-9.5)

Equation (1-9.5) can be put in its partial fraction as follows

$$\left(A + \frac{B}{(5s+9)}\right)ds = dx \qquad (1-9.6)$$

Where

$$A(5s+9) + B = (2s+5)$$
$$5A = 2 \qquad (1-9.7)$$
$$9A + B = 5$$

Thus,

$$A = \frac{2}{5}$$
$$B = \frac{7}{5} \qquad (1-9.8)$$

Therefore, equation (1-9.8) becomes

$$\int \left(\frac{2}{5} + \frac{7}{5(5s+9)}\right)ds = \int dx + C \qquad (1-9.9)$$

Integration gives

$$\frac{2}{5}s + \frac{7}{5}\ln(5s+9) = \ln x + C \qquad (1-9.10)$$

Substituting by $s = 2x + y$, from (1-9.2), we get

$$4x + 2y + 7\ln(10x + 5y + 9) = 5\ln x + C' \qquad (1-9.11)$$

1.4. Exact or Total first-order o.d.e.

The ordinary differential equation

$$f(x, y)dx + g(x, y)dy = 0 \qquad (1-10.1)$$

is total differential equation if $f(x,y)$ and $g(x,y)$ are independent partial derivatives of some function $h(x,y)$ such that

$$dh(x, y) = f(x, y)dx + g(x, y)dy = 0 \qquad (1-10.2)$$

This, upon integration, implies

24

$$h(x,y) = \text{constant} \tag{1-10.3}$$

Therefore, f(x,y) and g(x,y) are defined as

$$f(x,y) = \frac{\partial h(x,y)}{\partial x}$$
$$g(x,y) = \frac{\partial h(x,y)}{\partial y} \tag{1-10.4}$$

Equation (1-10.4) provides the nature of f and g when each is differentiated, f with respect to y, g to x, such that

$$\frac{\partial f(x,y)}{\partial y} = \frac{\partial g(x,y)}{\partial x} = \frac{\partial^2 h(x,y)}{\partial x \partial y} \tag{1-10.5}$$

Equation (1-10.5) provides the **necessary condition** for a first-order differential equation to exact or total.

Example 9

Solve the differential equation

$$(x+y+1)dx + (x - y^2 + 3)dy = 0 \tag{1-11.1}$$

Solution

The condition for a total derivative is

$$\frac{\partial f(x,y)}{\partial y} = \frac{\partial(x+y+1)}{\partial y} = 1$$
$$\frac{\partial g(x,y)}{\partial x} = \frac{\partial(x - y^2 + 3)}{\partial x} = 1 \tag{1-11.2}$$

Which suffices to prove that equation (1-11.1) is exact. Its solution requires determining h(x,y) = constant, as follows.

Equations (1-10.4) and (1-11.1) give

$$\frac{\partial h(x,y)}{\partial x} = x+y+1 \tag{1-11.3}$$

$$\frac{\partial h(x,y)}{\partial y} = x - y^2 + 3 \tag{1-11.4}$$

Integrating (1-11.3) with respect to x, we get an intermediate function in y as follows

$$h(x,y) = \int (x+y+1)dx$$
$$= \frac{x^2}{2} + xy + x + C(y) \tag{1-11.5}$$

From (1-11.4) and (1-11.5), we get the derivative of C(y) as follows

$$\frac{\partial h(x,y)}{\partial y} = x - y^2 + 3$$
$$= \frac{\partial}{\partial y}\left[\frac{x^2}{2} + xy + x + C(y)\right] \tag{1-11.6}$$
$$= x + \frac{dC(y)}{dy}$$

i.e.,

$$\frac{dC(y)}{dy} = -y^2 + 3 \tag{1-11.7}$$

Therefore, integration gives

$$C(y) = -\frac{y^3}{3} + 3y + k \tag{1-11.8}$$

Substituting by C(y), from (1-11.8) into (1-11.5), we get

$$h(x,y) = \frac{x^2}{2} + xy + x - \frac{y^3}{3} + 3y + k \tag{1-11.9}$$

This is equated to constant by virtue of (1-10.3),

$$3x^2 + 6xy + 6x - 2y^3 + 18y = K \tag{1-11.10}$$

Example 10

Solve the differential equation

$$\left(\frac{2x}{y^3}\right)dx + \left(\frac{y^2 - 3x^2}{y^4}\right)dy = 0 \tag{1-12.1}$$

Solution

26

The condition for a total derivative is

$$\frac{\partial f(x,y)}{\partial y} = \frac{\partial}{\partial y}\left(\frac{2x}{y^3}\right) = -\frac{6x}{y^4}$$

$$\frac{\partial g(x,y)}{\partial x} = \frac{\partial}{\partial x}\left(\frac{y^2 - 3x^2}{y^4}\right) = -\frac{6x}{y^4}$$

$$(1\text{-}12.2)$$

Which suffices to prove that equation (1-11.1) is exact. Its solution requires determining h(x,y) = constant, as follows.

Equations (1-10.4) and (1-12.1) give

$$\frac{\partial h(x,y)}{\partial x} = \frac{2x}{y^3} \qquad (1\text{-}12.3)$$

$$\frac{\partial h(x,y)}{\partial y} = \frac{y^2 - 3x^2}{y^4} \qquad (1\text{-}12.4)$$

Integrating (1-12.3) with respect to x, we get an intermediate function in y as follows

$$h(x,y) = \int \frac{2x}{y^3}\,dx$$

$$= \frac{x^2}{y^3} + C(y) \qquad (1\text{-}12.5)$$

From (1-12.4) and (1-12.5), we get the derivative of C(y) as follows

$$\frac{\partial h(x,y)}{\partial y} = \frac{y^2 - 3x^2}{y^4}$$

$$= \frac{\partial}{\partial y}\left[\frac{x^2}{y^3} + C(y)\right] \qquad (1\text{-}12.6)$$

$$= -3\frac{x^2}{y^4} + \frac{dC(y)}{dy}$$

i.e.,

$$\frac{dC(y)}{dy} = \frac{y^2 - 3x^2}{y^4} + 3\frac{x^2}{y^4} = \frac{1}{y^2} \qquad (1\text{-}12.7)$$

This upon integration gives

$$C(y) = -\frac{1}{y} + k \qquad\qquad (1\text{-}12.8)$$

Substituting by C(y), from (1-12.8) into (1-12.5), we get

$$h(x, y) = \frac{x^2}{y^3} - \frac{1}{y} + k \qquad\qquad (1\text{-}12.9)$$

Again, by virtue of (1-10.3), is equated to constant

$$\frac{x^2}{y^3} - \frac{1}{y} = K \qquad\qquad (1\text{-}12.10)$$

1.5. Integrating Factor or converting non-exact differential onto exact

If equation (1-10.1) is not exact, then necessary condition in equation (1-10.5) does not hold.

The **integration factor** is defined as an unknown function $\lambda(x,y)$ multiplier that converts equation (1-10.1) onto exact as follows

$$\lambda(x,y)f(x,y)dx + \lambda(x,y)g(x,y)dy = 0 \qquad\qquad (1\text{-}13.1)$$

Such that equation (1-10.5) is satisfied as follows

$$\frac{\partial}{\partial y}[\lambda(x,y)f(x,y)] = \frac{\partial}{\partial x}[\lambda(x,y)g(x,y)] \qquad\qquad (1\text{-}13.2)$$

Differentiating, we get

$$\lambda\frac{\partial f}{\partial y} + f\frac{\partial \lambda}{\partial y} = \lambda\frac{\partial g}{\partial x} + g\frac{\partial \lambda}{\partial x} \qquad\qquad (1\text{-}13.3)$$

Or

$$f\frac{\partial \lambda}{\partial y} - g\frac{\partial \lambda}{\partial x} = \lambda\left(\frac{\partial g}{\partial x} - \frac{\partial f}{\partial y}\right) \qquad\qquad (1\text{-}13.4)$$

Dividing both sides by $\lambda(x,y)$, we get

$$f\frac{1}{\lambda}\frac{\partial \lambda}{\partial y} - g\frac{1}{\lambda}\frac{\partial \lambda}{\partial x} = \frac{\partial g}{\partial x} - \frac{\partial f}{\partial y}$$

This can be put in the form

$$f \frac{\partial(\ln \lambda)}{\partial y} - g \frac{\partial(\ln \lambda)}{\partial x} = \frac{\partial g}{\partial x} - \frac{\partial f}{\partial y} \qquad (1\text{-}13.5)$$

Equation (1-13.5) determines the integrating function $\lambda(x,y)$ which converts the non-exact differential equation (1-10.1) onto exact.

Example 11

Solve the differential equation

$$\left(x^4 + x^2 y^2 + x\right)dx + ydy = 0 \qquad (1\text{-}14.1)$$

Solution

The condition for exact differential, equation (1-10.5) is not satisfied since

$$\frac{\partial}{\partial y}\left(x^4 + x^2 y^2 + x\right) \neq \frac{\partial}{\partial x} y \qquad (1\text{-}14.2)$$

$$2x^2 y \neq 0$$

Substituting by f and g from equation (1-14.1) into the condition for integrating factor, equation (1-13.5), we get

$$\left(x^4 + x^2 y^2 + x\right)\frac{\partial(\ln \lambda)}{\partial y} - y\frac{\partial(\ln \lambda)}{\partial x} = \frac{\partial}{\partial x} y - \frac{\partial}{\partial y}\left(x^4 + x^2 y^2 + x\right) \qquad (1\text{-}14.3)$$

$$= 0 - 2x^2 y$$

As we see, equation (1-14.3) is more complicated than the initial equation (1-14.1) without making the assumption that λ is independent of y.

Therefore, by assuming that λ is only function in x, equation (1-14.3) becomes

$$\frac{\partial(\ln \lambda)}{\partial x} = 2x^2 \qquad (1\text{-}14.4)$$

Integrating, we get

$$\lambda(x) = e^{\frac{2}{3}x^3} \qquad (1\text{-}14.5)$$

Multiplying equation (1-14.1) by $\lambda(x)$, we get an exact differential equation

$$\left(x^4 + x^2y^2 + x\right)e^{\frac{2}{3}x^3}dx + ye^{\frac{2}{3}x^3}dy = 0 \tag{1-14.6}$$

The exact function h(x,y) is easier obtained from the second term g(x,y), by equation (1-10.4), as follows

$$g(x,y) = \frac{\partial h(x,y)}{\partial y}$$

$$= ye^{\frac{2}{3}x^3} \tag{1-14.7}$$

Integrating, we get

$$h(x,y) = \int ye^{\frac{2}{3}x^3}dy$$

$$= \frac{y^2}{2}e^{\frac{2}{3}x^3} + C(x) \tag{1-14.8}$$

We can now retrace the derivatives of h, through equation (1-10.4), by differentiating (1-14.8) with respect to x and equating the derivative with the product $\lambda(x)f(x,y)$ from (1-14.6), as follows

$$\frac{\partial h(x,y)}{\partial x} = y^2x^2e^{\frac{2}{3}x^3} + \frac{dC(x)}{dx}$$

$$= \left(x^4 + x^2y^2 + x\right)e^{\frac{2}{3}x^3} \tag{1-14.9}$$

Therefore,

$$\frac{dC(x)}{dx} = \left(x^4 + x^2y^2 + x\right)e^{\frac{2}{3}x^3} - y^2x^2e^{\frac{2}{3}x^3}$$

$$= \left(x^4 + x\right)e^{\frac{2}{3}x^3} \tag{1-14.10}$$

Integrating, we get

$$C(x) = \int \left(x^4 + x\right)e^{\frac{2}{3}x^3}dx \tag{1-14.11}$$

Integrating by parts, we get

$$C(x) = \int x^4 e^{\frac{2}{3}x^3} dx + \int xe^{\frac{2}{3}x^3} dx$$
$$= \frac{1}{2}x^2 e^{\frac{2}{3}x^3} - \int e^{\frac{2}{3}x^3} xdx + \int xe^{\frac{2}{3}x^3} dx \qquad (1\text{-}14.12)$$
$$= \frac{1}{2}x^2 e^{\frac{2}{3}x^3}$$

Substituting from equation (1-14.12) in (1-14.8), we get

$$h(x,y) = \frac{y^2}{2}e^{\frac{2}{3}x^3} + \frac{x^2}{2}e^{\frac{2}{3}x^3} \qquad (1\text{-}14.13)$$

By equation (1-10.3), the exact function becomes

$$\left(x^2 + y^2\right)e^{\frac{2}{3}x^3} = K \qquad (1\text{-}14.14)$$

Proof checking

$$f(x,y) = \frac{\partial h(x,y)}{\partial x}$$
$$= \frac{1}{2}\frac{\partial}{\partial x}\left(x^2 + y^2\right)e^{\frac{2}{3}x^3} \qquad (1\text{-}14.15)$$
$$= \left(x + x^4 + y^2 x^2\right)e^{\frac{2}{3}x^3}$$

Which yields the $\lambda(x,y)f(x,y)$, equation (1-14.1) and (1-14.6).

And

$$g(x,y) = \frac{\partial h(x,y)}{\partial y}$$
$$= \frac{1}{2}\frac{\partial}{\partial y}\left(x^2 + y^2\right)e^{\frac{2}{3}x^3} \qquad (1\text{-}14.16)$$
$$= ye^{\frac{2}{3}x^3}$$

Which yields the $\lambda(x,y)g(x,y)$, equation (1-14.1) and (1-14.6).

Thus, the integrating factor converted the incomplete differential function (1-14.1) into the complete differential function (1-14.6).

Example 12

31

Solve the differential equation

$$(y + xy^2)dx - xdy = 0 \tag{1-15.1}$$

Solution

The condition for exact differential, equation (1-10.5) is not satisfied since

$$\frac{\partial}{\partial y}(y + xy^2) \neq \frac{\partial}{\partial x}(-x) \tag{1-15.2}$$

$$1 + 2xy \neq -1$$

Substituting by f and g from equation (1-15.1) into the condition for integrating factor, equation (1-13.5), we get

$$(y + xy^2)\frac{\partial(\ln \lambda)}{\partial y} + x\frac{\partial(\ln \lambda)}{\partial x} = \frac{\partial}{\partial x}(-x) - \frac{\partial}{\partial y}(y + xy^2) \tag{1-15.3}$$

$$= -2 - 2xy$$

Which can be written in simple form as follows

$$y(1 + xy)\frac{\partial(\ln \lambda)}{\partial y} + x\frac{\partial(\ln \lambda)}{\partial x} = -2(1 + xy) \tag{1-15.4}$$

As we see, equation (1-15.4) permits us finding an integrator factor $\lambda(y)$ independent of x, which will eliminate the derivative of λ with respect to x. Thus, dividing by (1+xy), we get

$$d(\ln \lambda) = -2\frac{dy}{y} \tag{1-15.5}$$

Integrating, we get

$$\ln \lambda = -2 \ln y$$

$$\lambda = y^{-2} \tag{1-15.6}$$

Multiplying equation (1-15.1) by $\lambda(x)$, we get an exact differential equation

$$\frac{(y + xy^2)}{y^2}dx - \frac{x}{y^2}dy = 0 \tag{1-15.7}$$

The exact function h(x,y) is easier obtained from the second term g(x,y), by equation (1-10.4), as follows

$$g(x,y) = \frac{\partial h(x,y)}{\partial y}$$

$$= -\frac{x}{y^2}$$

(1-15.8)

Integrating, we get

$$h(x,y) = \int -\frac{x}{y^2} dy$$

$$= \frac{x}{y} + C(x)$$

(1-15.9)

We can now retrace the derivatives of h, through equation (1-10.4), by differentiating (1-15.9) with respect to x and equating the derivative with the product $\lambda(y)f(x,y)$ from (1-15.7), as follows

$$\frac{\partial h(x,y)}{\partial x} = \frac{1}{y} + \frac{dC(x)}{dx}$$

$$= \left(\frac{y + xy^2}{y^2} \right)$$

(1-15.10)

Therefore,

$$\frac{dC(x)}{dx} = x$$

(1-15.11)

Integrating, we get

$$C(x) = \frac{x^2}{2}$$

(1-15.12)

Substituting from equation (1-15.12) in (1-15.9), we get

$$h(x,y) = \frac{x}{y} + \frac{x^2}{2}$$

(1-15.13)

By equation (1-10.3), the exact function becomes

$$\frac{x}{y} + \frac{x^2}{2} = K$$

(1-15.14)

1.6. Linear First-Order o.d.e.

33

The first order o.d.e. is linear if its derivative and unknown functions are related by linear operations of the form

$$\frac{dy}{dx} + P(x)y = Q(x) \qquad (1\text{-}16.1)$$

Where $P(x)$ and $Q(x)$ are continuous functions in x and the derivative of y is of first order.

1.6.1. Solution of linear first-order o.d.e. by integrating factor

Writing equation (1-16.1) by unknown integrating factor $\lambda(x)$ in the form

$$(\lambda Py - \lambda Q)dx + \lambda dy = 0 \qquad (1\text{-}16.2)$$

Where,

$$f(x,y) = \lambda Py - \lambda Q$$
$$g(x,y) = \lambda dy \qquad (1\text{-}16.3)$$

The condition for exact differential equation, given in (1-10.5), implies

$$\frac{\partial f(x,y)}{\partial y} = \frac{\partial g(x,y)}{\partial x}$$
$$\frac{\partial(\lambda Py - \lambda Q)}{\partial y} = \frac{\partial \lambda}{\partial x} \qquad (1\text{-}16.4)$$

Noting that P, Q, and λ are independent of x, the derivatives of (1-16.4) gives

$$\lambda P = \frac{d\lambda}{dx} \qquad (1\text{-}16.5)$$

Integrating, we get

$$\frac{d\lambda}{\lambda} = Pdx \qquad (1\text{-}16.6a)$$

Thus, the Integrating factor is given by

$$\lambda = e^{\int Pdx} \qquad (1\text{-}16.6b)$$

Substituting from (1-16.6) into (1-16.2), we get

$$yd\lambda - \lambda Qdx + \lambda dy = 0 \qquad (1\text{-}16.7)$$

34

This can be written concisely in the form

$$d(y\lambda) = \lambda Q dx \qquad (1\text{-}16.8)$$

Upon integrating, we get

$$y\lambda = \int \lambda Q dx + C \qquad (1\text{-}16.9)$$

Summary

The separable 1st-order o.d.e.	$\dfrac{dy}{dx} + P(x)y = Q(x)$	(1-16.1)
Integrating Factor:	$\lambda = e^{\int Pdx}$	(1-16.6b)
Solution of 1st-order o.d.e.	$y = \dfrac{\int e^{\int Pdx} Q dx + C}{e^{\int Pdx}}$	(1-16.9)

Example 13

Solve the differential equation

$$(x+1)\frac{dy}{dx} - 2y = (x+1)^4 \qquad (1\text{-}17.1)$$

Solution

Equation (1-17.1) is divided by $(x+1)$ to take the linear first order o.d.e. as

$$\frac{dy}{dx} - \frac{2y}{x+1} = (x+1)^3 \qquad (1\text{-}17.2)$$

Equation (1-17.2) is linear because both y and its derivative are first order. The P(x) and Q(x) functions defined in (1-16.2) are given explicitly in equation (1-17.2) as follows

$$P(x) = -\frac{2}{x+1}$$
$$Q(x) = (x+1)^3 \qquad (1\text{-}17.3)$$

The integrating factor given by equation (1-16.6) and (1-17.3) as follows

35

$$\frac{d\lambda}{\lambda} = P(x)dx$$

$$= -\frac{2}{x+1}dx \qquad (1\text{-}17.4)$$

Integrating, we get

$$\lambda = e^{\int -\frac{2}{x+1}dx}$$

$$= e^{-2\ln(x+1)}$$

Or

$$\lambda = \frac{1}{(x+1)^2} \qquad (1\text{-}17.5)$$

The solution of equation (1-17.1), is given in terms of $\lambda(x)$ by equation (1-16.9), with substitution from (1-17.3), as follows

$$y\lambda = \int \lambda Q dx + C$$

$$y\frac{1}{(x+1)^2} = \int \frac{1}{(x+1)^2}(x+1)^3 dx + C \qquad (1\text{-}17.6)$$

$$= \frac{x^2}{2} + x + C$$

Thus, the final solution of equation (1-17.1) is

$$y = \left(\frac{x^2}{2} + x + C\right)(x+1)^2 \qquad (1\text{-}17.7)$$

Example 14

Solve the differential equation

$$\frac{dy}{dx} - y\cot x = 2x\sin x \qquad (1\text{-}18.1)$$

Solution

Equation (1-18.2) is linear because both y and its derivative are first order. The P(x) and Q(x) functions defined in (1-16.2) are given explicitly in equation (1-18.1) as follows

$$P(x) = -\cot x$$
$$Q(x) = 2x \sin x$$
(1-18.2)

The integrating factor given by equation (1-16.6) and (1-18.2) as follows

$$\frac{d\lambda}{\lambda} = P(x)dx$$
(1-18.3)

$$= -\cot x dx$$

Integrating, we get

$$\lambda = e^{\int -\cot x dx}$$

$$= e^{-\ln \sin x}$$

Or

$$\lambda = \frac{1}{\sin x}$$
(1-18.4)

The solution of equation (1-18.1), is given in terms of $\lambda(x)$ by equation (1-16.9), with substitution from (1-18.2), as follows

$$y\lambda = \int \lambda Q dx + C$$
$$y \frac{1}{\sin x} = \int \frac{2x}{\sin x} \sin x dx + C$$
(1-18.5)
$$= x^2 + C$$

Thus, the final solution of equation (1-18.1) is

$$y = (x^2 + C)\sin x$$
(1-18.6)

1.6.2. Solution of linear first-order o.d.e. by a product of two functions

The linear first-order o.d.e. (1-16.1) can be solved by a product of two functions, such that one of the two is set to simplify the problem by eliminating complex terms, as follows.

Given the differential equation

$$\frac{dy}{dx} + P(x)y = Q(x)$$
(1-19.1)

With a proposed solution of the form

$$y = f(x)g(x) \qquad (1\text{-}19.2)$$

Differentiating (1-19.2), we get

$$\frac{dy}{dx} = f(x)\frac{dg(x)}{dx} + g(x)\frac{df(x)}{dx} \qquad (1\text{-}19.3)$$

Substituting from (1-19.3) in (1-19.1), we get

$$f\frac{dg}{dx} + g\frac{df}{dx} + Pfg = Q \qquad (1\text{-}19.4)$$

Where we dropped the bracketed arguments to simplify notation.

Equation (1-19.4) can be written as follows:

$$f\frac{dg}{dx} + g\left(\frac{df}{dx} + Pf\right) = Q \qquad (1\text{-}19.5)$$

Equation (1-19.5) represents the main formula where the present method is founded, which is to equate the bracketed term to zero and which determines one of the two unknown functions, g and f.

Therefore,

$$\frac{df}{dx} + Pf = 0 \qquad (1\text{-}19.6)$$

Equation (1-19.6) determines f(x) by integration and simplifies equation (1-19.5) to give

$$f\frac{dg}{dx} = Q \qquad (1\text{-}19.7)$$

Integrating (1-19.6), we get

$$\int \frac{df}{f} = -\int Pdx + C$$
$$f(x) = Ce^{-\int Pdx} \qquad (1\text{-}19.8)$$

Hence, we determine one of the two functions f and g, of (1-19.2). We then substitute by f from (1-19.8) into equation (1-19.7), and integrate to obtain g(x), as follows

$$f\frac{dg}{dx} = Q$$

$$Ce^{-\int Pdx}\frac{dg}{dx} = Q \tag{1-19.9}$$

Therefore,

$$g(x) = \int \frac{Q(x)}{C}e^{\int Pdx}dx \tag{1-19.10}$$

Thus, the final solution of (1-19.1), is

$$y = f(x)g(x)$$
$$= e^{-\int Pdx}\left[\int Q(x)e^{\int Pdx}dx\right] \tag{1-19.11}$$

Example 15

Solve the differential equation

$$\frac{dy}{dx} - \frac{y}{x} = x^2 \tag{1-20.1}$$

Solution

Assume a solution of the form y=f(x)g(x), equation (1-19.2). Equation (1-20.1) becomes

$$g\frac{df}{dx} + f\frac{dg}{dx} - \frac{gf}{x} = x^2 \tag{1-20.2}$$

i.e.,

$$g\frac{df}{dx} + f\left(\frac{dg}{dx} - \frac{g}{x}\right) = x^2 \tag{1-20.3}$$

Equate the bracketed term with zero

$$\frac{dg}{dx} - \frac{g}{x} = 0 \tag{1-20.4}$$

Arranging and integrating. we get

$$\int \frac{dg}{g} = \int \frac{dx}{x} \tag{1-20.4}$$
$$\ln g = \ln x + C$$

i.e., $\qquad g = x + C'$ (1-20.5)

Substitute by g from (1-20.5) into equation (1-20.3), and noting the vanishing bracket, we get

$$(x + C')\frac{df}{dx} = x^2$$ (1-20.6)

Or,

$$df = \frac{x^2}{x + C'}dx$$ (1-20.7)

Since C' is an arbitrary constant, the integration of (1-20.7) gives

$$f = \frac{1}{2}x^2 + C$$ (1-20.8)

Hence, the solution of (1-20.1) is

$$y = gf = \frac{1}{2}x^3 + xC$$ (1-20.9)

1-7. Linearization of Bernoulli's Equation

This is a nonlinear equation, which can be reduced to linear form and has the following form:

$$\frac{dy}{dx} + P(x)y = Q(x)y^n$$ (1-21.1)

Where n has any real value other than zero or one ($n \neq 0$ and $n \neq 1$).

Divide equation (1-21.1) by y^n, we get

$$y^{-n}\frac{dy}{dx} + P(x)y^{1-n} = Q(x)$$ (1-21.2)

Substitute by

$$s = y^{-n+1}$$ (1-21.3)

With the derivative

$$\frac{ds}{dx} = (-n+1)y^{-n}\frac{dy}{dx} \tag{1-21.4}$$

Substituting by s and its derivative, equations (1-21.3) and (1-21.4) in (1-21.2), we get

$$y^{-n}\frac{dy}{dx} + P(x)y^{1-n} = Q(x)$$

$$\frac{ds}{dx} + (-n+1)P(x)s = (-n+1)Q(x) \tag{1-21.5}$$

Equation (1-21.5) is the linear transformation of Bernoulli's equation.

Example 16

Solve the differential equation

$$\frac{dy}{dx} + yx = x^3y^3 \tag{1-22.1}$$

Solution

Divide equation (1-22.1) by y^3, we get

$$y^{-3}\frac{dy}{dx} + xy^{-2} = x^3 \tag{1-22.2}$$

Substitute by

$$s = y^{-2} \tag{1-22.3}$$

With the derivative

$$\frac{ds}{dx} = -2y^{-3}\frac{dy}{dx} \tag{1-22.4}$$

Substituting by s and its derivative, equations (1-22.3) and (1-22.4) in (1-22.2), we get

$$\frac{ds}{dx} - 2xs = -2x^3 \tag{1-22.5}$$

Equation (1-22.5) is a first-order linear o.d.e., which solution takes the form

41

$$s = fg$$

$$\frac{ds}{dx} = g\frac{df}{dx} + f\frac{dg}{dx} \tag{1-22.6}$$

Substituting by s and its derivative, from (1-22.6) into (1-22.5), we get

$$g\frac{df}{dx} + f\frac{dg}{dx} - 2xfg = -2x^3 \tag{1-22.7}$$

Lumping terms that could cancel in order to determine one of the two arbitrary functions f and g, we get

$$g\frac{df}{dx} + f\left(\frac{dg}{dx} - 2xg\right) = -2x^3 \tag{1-22.7}$$

Hence, we set

$$\frac{dg}{dx} - 2xg = 0 \tag{1-22.8}$$

And, therefore,

$$g\frac{df}{dx} = -2x^3 \tag{1-22.9}$$

From equation (1-22.8), integration gives

$$\int \frac{dg}{g} = \int 2xdx \tag{1-22.10}$$

Or,

$$g = Ce^{x^2} \tag{1-22.11}$$

Substituting by g, equation (1-22.11) in (1-22.9), and manipulating the arbitrary constant to ease the formulation, we get

$$Ce^{x^2}\frac{df}{dx} = -2x^3$$

$$\frac{df}{dx} = C'\left(-2x^3e^{-x^2}\right)$$

(1-22.12)

C' is a constant.

Integration by parts, we get

$$f = \int -2x^3e^{-x^2}dx + C$$

$$= \int x^2de^{-x^2} + C$$

$$= x^2e^{-x^2} - \int e^{-x^2}dx^2 + C$$

$$= x^2e^{-x^2} + e^{-x^2} + C$$

(1-22.13)

From (1-22.11) and (1-22.13), the solution of (1-22.5) is

$$s = gf$$

$$= e^{x^2}\left(x^2e^{-x^2} + e^{-x^2} + C\right)$$

$$= x^2 + 1 + Ce^{x^2}$$

(1-22.14)

Therefore, the solution, y, for Bernoulli's equation, is obtained from (1-22.14) and (1-22.3) as

$$s = y^{-2}$$

$$= x^2 + 1 + Ce^{x^2}$$

(1-22.15)

i.e.,

$$y = \frac{1}{\sqrt{x^2 + 1 + Ce^{x^2}}}$$

(1-22.15)

1-8. Linearization of Riccati's Equation

This is a nonlinear equation, which can be transformed to Bernoulli's equation, then reduced to linear form.

Riccati's equation has the form

$$\frac{dy}{dx} + P(x)y + Q(x)y^2 = R(x) \tag{1-23.1}$$

Assume that s(x) is a particular solution of (1-23.1).

Substitute by

$$y(x) = s(x) + z(x) \tag{1-23.2}$$

Where z(x) is an unknown function to be determined.

Substituting by y and its derivatives from (1-23.2) into (1-23.1), we get

$$\frac{ds}{dx} + \frac{dz}{dx} + P(x)(s+z) + Q(x)(s+z)^2 = R(x) \tag{1-23.3}$$

Since s(x) is assumed a particular solution of (1-23.1), we could eliminate R(x), from (1-23.3) by expanding (1-23.3) and using (1-23.1) to get the following two equations:

$$\frac{ds}{dx} + P(x)s + Q(x)s^2 = R(x) \tag{1-23.4}$$

$$\frac{dz}{dx} + P(x)z + Q(x)(2sz + z^2) = 0 \tag{1-23.5}$$

Equation (1-23.5) can be arranged to take the form of Bernoulli's equation as follows

$$\frac{dz}{dx} + [P(x) + 2Q(x)]z = -Q(x)z^2 \tag{1-23.6}$$

Example 17

Solve the differential equation

$$\frac{dy}{dx} = y^2 - \frac{2}{x^2}$$

(1-24.1)

Solution

Due to the structure of the RHS of equation (1-24.1), it could have a particular solution of the form

$$s = \frac{C}{x}$$

(1-24.2)

Which renders the derivative y' vanishing.

Thus, a general solution of the form y(x) = s(x) + z(x), equation (1-23.2), takes the form

$$y = z + \frac{1}{x}$$

(1-24.3)

Differentiating, we get

$$\frac{dy}{dx} = \frac{dz}{dx} - \frac{1}{x^2}$$

(1-24.4)

Substituting by y' from (1-24.1) and by y from (1-24.3), we get

$$\frac{dy}{dx} = \frac{dz}{dx} - \frac{1}{x^2}$$
$$\left(z + \frac{1}{x}\right)^2 - \frac{2}{x^2} = \frac{dz}{dx} - \frac{1}{x^2}$$

(1-24.5)

Expanding and arranging, we get

$$\frac{dz}{dx} - \frac{2z}{x} = z^2$$

(1-24.5)

Equation (1-24.5) is of the Bernoulli's type, (1-21.1) because it is first-order o.d.e. and entails first degree and second-degree terms of dependent variables z.

Substitute in (1-24.5) by

$$u = z^{-1}$$

$$\frac{du}{dz} = -\frac{1}{z^2}\frac{du}{dz}$$

(1-24.6)

We get the linear o.d.e.

$$\frac{du}{dx} + \frac{2u}{x} = -1$$

(1-24.7)

The integrating factor of (1-24.7) is obtained by putting it in the form (1-13.1) and applying the condition for completing to the exact differential equation as follows

$$\lambda(2u + x)dx + \lambda x du = 0$$

(1-24.8)

From equation (1-16.2), we get

$$P(x) = \frac{2}{x}$$
$$Q(x) = -1$$

(1-24.9)

Substituting from (1-24.9) in the (1-16.6), we get

$$\frac{d\lambda}{\lambda} = \frac{2}{x}dx$$

(1-24.10)

Integrating, we get

$$\ln \lambda = 2\ln x$$

(1-24.11)

Integrating, we get

$$\lambda = x^2$$

(1-24.12)

Therefore, the solution of equation (1-24.7) is obtained by (1-16.9) and (1-24.9) as follows

$$u\lambda = \int \lambda Q dx + C$$

$$= -\int x^2 dx + C$$

$$= -\frac{1}{3}x^3 + C$$

(1-24.13)

Or,

46

$$u = -\frac{1}{3}x + \frac{1}{x^2}C \qquad (1\text{-}24.14)$$

Substituting by u from (1-24.14) in (1-24.6), we get

$$z = \frac{3x^2}{-x^2 + 3C} \qquad (1\text{-}24.15)$$

Substituting by z from (1-24.15) in (1-24.3), we get

$$y = \frac{3x^2}{-x^2 + C'} + \frac{1}{x} \qquad (1\text{-}24.16)$$

C' is a constant.

1-9. Exercises on first-order first-degree o.d.e.

Solve the following o.d.e.

1-1. $\tan y\, dx - \cot x\, dy = 0$ [Ans: $\sin y \cos x = C$]

1-2. $(12x + 5y - 9)\, dx + (5x + 2y - 3)dy = 0$ [Ans: $6x^2 + 5xy + y^2 - 9x - 3y = C$]

1-3. $x\dfrac{dy}{dx} + y = x^3$ $\left[y = \dfrac{C}{x} + \dfrac{1}{4}x^3 \right]$

1-4. $y\,dx - x\,dy = x^2 y\,dy$ $\left[\text{Ans}: \dfrac{y^2}{2} + \dfrac{y}{x} = C \right]$

1-5. $\dfrac{dx}{dt} + 3x = e^{2t}$ $\left[\text{Ans}: \ x = Ce^{-3t} + \dfrac{1}{5}e^{2t} \right]$

1-6. $y \sin x + y' \cos x = 1$ $[\, y = C \cos x + \sin x \,]$

1-7. $\dfrac{dx}{dt} = x + \sin t$ $\left[\text{Ans}: \ x = Ce^t - \dfrac{1}{2}(\cos t + \sin t) \right]$

1-8. $x(\ln x - \ln y)dy - y\,dx = 0$ $\left[\text{Ans}: \ x = ye^{cy+1} \right]$

1-9. $\dfrac{dx}{dt} = e^{x/t} + \dfrac{x}{t}$ $\left[\text{Ans}: \ \ln t = C - e^{-x/t} \right]$

1-10. $y = x\dfrac{dy}{dx} + \dfrac{1}{y}$ $\left[\text{Ans}: \ y = Cx + \dfrac{1}{C} \right]$

1-11. $(2x + 2y - 1)dx + (x + y - 2)dy = 0$ $\left[\text{Ans}: \ (x+y+1)^3 = Ce^{2x+y} \right]$

1-12. $(x - y)ydx - x^2dy = 0$ $\left[\text{Ans}: \ x = Ce^{\frac{x}{y}} \right]$

1-13. $\dfrac{dy}{dx} = \dfrac{3x - 4y - 2}{3x - 4y - 3}$ $\left[\text{Ans}: \ 3x - 4y + 1 = e^{xy} \right]$

1-14. $\dfrac{dx}{dt} - x\cot t = 4\sin t$ $\left[\text{Ans}: \ x = (4t + C)\sin t \right]$

1-15. $\dfrac{dy}{dx} - \dfrac{3y}{x} + x^3y^2 = 0$ $\left[\text{Ans}: \ y = \dfrac{7x^3}{x^7 + C} \right]$

1-16. $(x^2 - y)dx + (x^2y^2 + x)dy = 0$ $\left[\text{Ans}: \ 3(x^2 + y) + xy^3 = Cx \right]$

1-17. Find the **integrating factor** of the equation

$(3y^2 - x)dx + 2y(y^2 - 3x)dy = 0$ $\left[\text{Ans}: \ \text{I.F.} = \dfrac{C}{(x + y^2)^3} \right]$

1-18. $(x - y)ydx - x^2dy = 0$ $\left[\text{Ans}: \ x = Ce^{x/y} \right]$

1-19. $\dfrac{dy}{dx} = \dfrac{x + y - 3}{1 - x + y}$ $\left[\text{Ans}: \ (x^2 + 2xy - y^2 - 6x - 2y = C \right]$

1-20. $x\dfrac{dy}{dx} - y^2 \ln x + y = 0$ $\left[\text{Ans}: \ y = \dfrac{1}{1 + Cx + \ln x} \right]$

1-21. $(x^2 - 1)\dfrac{dy}{dx} + 2xy - \cos x = 0$ $\left[\text{Ans}: \ (x^2 - 1)y - \sin x = C \right]$

1-22. $(4y + 2x + 3)\dfrac{dy}{dx} - 2y - x - 1 = 0$ $\left[\text{Ans}: \ 8y + 4x + 5 = Ce^{4x - 8y - 4} \right]$

1-23. $(y^2 - x)\dfrac{dy}{dx} - y + x^2 = 0$ $\left[\text{Ans}: \ x^3 + y^3 - 3xy = C \right]$

1-24. $(y^2 - x^2)\dfrac{dy}{dx} + 2xy = 0$ $\left[\text{Ans}: \ y = C(x^2 + y^2) \right]$

1-25. $3xy^2\dfrac{dy}{dx} + y^3 - 2x = 0$ $\left[\text{Ans}: \ y^3 = x + \dfrac{C}{x} \right]$

1-26. $xy^2dy = (x^3 + y^3)dx$ $\left[\text{Ans}: \ y = x(3\ln Cx)^{1/3} \right]$

1-27. $\dfrac{dy}{dx} + xy = x^3y^3$ $\left[\text{Ans}: \ y^2(x^2 + 1 + Ce^{x^2}) = 1 \right]$

1-28. $y - \cos x \dfrac{dy}{dx} = y^2 \cos x (1 - \sin x)$ $\left[\text{Ans}: \quad y = \dfrac{\tan x - \sec x}{\sin x + C} \right]$

1-29. $xy \dfrac{dy}{dx} = 1 - x^2$ $\left[\text{Ans}: \quad x^2 + y^2 \ln C x^2 \right]$

1-30. $(1 + e^x) y \dfrac{dy}{dx} = e^x$

        $y = 1 \quad \text{when} \quad x = 0$ $\left[\text{Ans}: \quad 2 e^{\frac{1}{2} y^2} = \sqrt{e}(1 + e^x) \right]$

1-31. $(2x + 3y - 1)dx + (4x + 6y - 5)dy = 0$ $\left[\text{Ans}: \quad x + 2y + 3 \ln(2x + 3y - 7) = C \right]$

1-32. $(4x^2 + 3xy + y^2)dx + (4y^2 + 3xy + x^2)dy = 0$ $\left[\text{Ans}: \quad (x^2 + y^2)^3 (x + y)^2 = C \right]$

1-33. $(1 + x + y) \dfrac{dy}{dx} = 1 - 3x - 3y$ $\left[\text{Ans}: \quad 3x + y + 2 \ln(x + y - 1) = C \right]$

1-34. $y(1 + xy)dx - x dy = 0$ $\left[\text{Ans}: \quad \dfrac{x}{y} + \dfrac{x^2}{2} = C \right]$

1-35. $\dfrac{y}{x} dx + (y^3 - \ln x)dy = 0$ $\left[\text{Ans}: \quad \dfrac{1}{y} \ln x + \dfrac{y^2}{2} = C \right]$

CHAPTER 2

FIRST-ORDER O.D.E. OF HIGHER DEGREES

2-1. Solution for derivatives

Example 18

Solve the differential equation

$$\left(\frac{dy}{dx}\right)^2 - (x+y)\frac{dy}{dx} + xy = 0 \tag{2-1.1}$$

Solution

Equation (2-1) has the algebraic form

$$\left(\frac{dy}{dx} - x\right)\left(\frac{dy}{dx} - y\right) = 0 \tag{2-1.2}$$

We thus have two integrable equations

$$\frac{dy}{dx} - x = 0 \tag{2-1.3}$$

$$\frac{dy}{dx} - y = 0 \tag{2-1.4}$$

Integrating (2-1.3) yields

$$\int dy = \int xdx$$
$$y = \frac{x^2}{2} + C_2 \qquad (2\text{-}1.5)$$

Integrating (2-1.4) yields
$$\ln y = x + \ln C_1$$
$$y = C_1 e^x \qquad (2\text{-}1.6)$$

The two solutions (2-1.5) and (2-1.6) give the final solution

$$\left(y - \frac{x^2}{2} + C_2\right)\left(y - C_1 e^x\right) = 0 \qquad (2\text{-}1.7)$$

Example 19

Solve the differential equation

$$\left(\frac{dy}{dx}\right)^3 - e^{2x}\frac{dy}{dx} = 0 \qquad (2\text{-}2.1)$$

Solution

Equation (2-2.1) has the algebraic form

$$\frac{dy}{dx}\left(\left(\frac{dy}{dx}\right)^2 - e^{2x}\right) = 0 \qquad (2\text{-}2.2)$$

We thus have two integrable equations

$$\frac{dy}{dx} = 0 \qquad (2\text{-}2.3)$$

$$\frac{dy}{dx} = \sqrt{e^{2x}} = \pm e^x \qquad (2\text{-}2.4)$$

Integrating (2-2.3) yields

$$y = C_1 \qquad (2\text{-}2.5)$$

Integrating (2-2.4) yields

$$y = \pm e^x + C_2 \tag{2-2.6}$$

The two solutions (2-2.5) and (2-2.6) give the final solution

$$\left(y - e^x + C_2\right)\left(y + e^x + C_2\right)\left(y + C_1\right) = 0 \tag{2-2.7}$$

Example 20

Solve the differential equation

$$\left(\frac{dy}{dx}\right)^2 + 2y\frac{dy}{dx}\cot x - y^2 = 0 \tag{2-3.1}$$

Solution

Equation (2-3.1) is quadratic in the derivative of y and has two solutions given by

$$\frac{dy}{dx} = -y\cot x \pm \sqrt{y^2 \cot^2 x + y^2} \tag{2-3.2}$$

$$= y\left(-\cot x \pm \operatorname{cosec} x\right)$$

We thus have two integrable equations

$$\int \frac{dy}{y} = \int \left(-\cot x \pm \operatorname{cosec} x\right) dx \tag{2-3.3}$$

Integrating, we get

$$\ln y = \ln \operatorname{cosec} x \pm \ln(\operatorname{cosec} x - \cot x) + \ln C \tag{2-3.4}$$

This can be written as

$$y_1 = C \operatorname{cosec} x(\operatorname{cosec} x - \cot x)$$
$$= C\frac{1}{\sin x}\left(\frac{1}{\sin x} - \frac{\cos x}{\sin x}\right)$$
$$= C\left(\frac{1 - \cos x}{\sin^2 x}\right) \tag{2-3.5}$$
$$= \frac{C}{1 + \cos x}$$

$$y_2 = C\frac{\operatorname{cosec} x}{(\operatorname{cosec} x - \cot x)}$$

$$= C\frac{1}{\sin x\left(\dfrac{1}{\sin x} - \dfrac{\cos x}{\sin x}\right)} \tag{2-3.6}$$

$$= C\frac{1}{1 - \cos x}$$

The two solutions (2-3.5) and (2-3.6) give the final solution

$$\left(y - \frac{C}{1 - \cos x}\right)\left(y - \frac{C}{1 + \cos x}\right) = 0 \tag{2-3.7}$$

2-2. Solution for dependent variable

Example 21

Solve the differential equation

$$\left(\frac{dy}{dx}\right)^2 - 2x\frac{dy}{dx} + y = 0 \tag{2-4.1}$$

Solution

Equation (2-4.1) is readily solvable for y such that

$$y = 2x\frac{dy}{dx} - \left(\frac{dy}{dx}\right)^2 \tag{2-4.2}$$

Differentiating with respect to x, we get

$$\frac{dy}{dx} = 2\frac{dy}{dx} + 2x\frac{d^2y}{dx^2} - 2\left(\frac{dy}{dx}\right)\frac{d^2y}{dx^2} \tag{2-4.3}$$

Arranging, we get

$$\frac{d^2y}{dx^2} = \frac{\dfrac{dy}{dx}}{2\left(\dfrac{dy}{dx} - x\right)} \qquad (2\text{-}4.4)$$

Equation (2-4.4) can be homogenized to the form of equation (1-4.1) by the two substitutions:

$$P = \frac{dy}{dx}, \qquad \frac{dP}{dx} = \frac{d^2y}{dx^2}, \qquad s = \frac{\dfrac{dy}{dx}}{x} = \frac{P}{x} \qquad (2\text{-}4.5)$$

Differentiating s we get

$$\frac{ds}{dx} = \frac{x\dfrac{dP}{dx} - P}{x^2} = \frac{\dfrac{dP}{dx} - s}{x} \qquad (2\text{-}4.6)$$

$$\frac{dP}{dx} = x\frac{ds}{dx} + s$$

Substituting from equations (2-4.5) and (2-4.6, we get

$$\frac{d^2y}{dx^2} = \frac{\dfrac{dy}{dx}}{2\left(\dfrac{dy}{dx} - x\right)} \qquad (2\text{-}4.7)$$

$$x\frac{ds}{dx} + s = \frac{s}{2(s-1)}$$

Arranging, we get

$$x\frac{ds}{dx} = \frac{-(2s-3)s}{2(s-1)} \qquad (2\text{-}4.8)$$

Equation (2-4.8) is written in integrable form as follows

$$\frac{dx}{x} = -\frac{2(s-1)}{(2s-3)s}\,ds$$

$$= \left[\frac{B}{s} + \frac{A}{(2s-3)}\right]ds \qquad (2\text{-}4.9)$$

54

Where the constant A and B are determined from the coefficients of terms of equal powers in the equations

$$B(2s - 3) + As = -2(s - 1)$$
$$2B + A = -2$$
$$B = -\frac{2}{3} \qquad (2\text{-}4.10)$$
$$A = -\frac{2}{3}$$

Substituting by A and B in (2-4.9), we get

$$\int \frac{dx}{x} = -\frac{2}{3} \int \left[\frac{1}{s} + \frac{1}{(2s - 3)}\right] ds \qquad (2\text{-}4.11)$$

Integrating, we get

$$\ln x = -\frac{2}{3}\left[\ln s + \frac{1}{2}\ln(2s - 3)\right] + \ln C \qquad (2\text{-}4.12)$$

Thus,

$$x^3 = \frac{C}{\left(s^2(2s - 3)\right)} \qquad (2\text{-}4.13)$$

i.e.,
$$x^3 s^2 (2s - 3) = C \qquad (2\text{-}4.14)$$

Retracing our substitutions from (2-4.5), equation (2-4.14) becomes

i.e.,
$$P^2(2P - 3x) = C \qquad (2\text{-}4.15)$$

We have arrived at the expression of x in terms of P by arranging the above equation as follows

i.e.,
$$x = \frac{2}{3}P + \frac{C}{P^2} \qquad (2\text{-}4.16)$$

We note the redundancy in changing the constant C to absorb numerical values and signs such that he negative sign and division by 3 in the above equation.

Substituting by x, from the above equation, in (2-4.2), and noting that P is the derivative of y, (2-4.5), we get

$$y = 2\left(\frac{2}{3}P + \frac{C}{P^2}\right)P - P^2$$

$$= \frac{1}{3}P^2 + \frac{2C}{P}$$

(2-4.17)

Summary of solution

The parametric representation of **complete integral** (parametric equations of general solution) are given by equations (2-4.16) and 92-4.17) as follows

$$x = \frac{2}{3}P + \frac{C}{P^2}$$

(2-4.16)

$$y = \frac{1}{3}P^2 + \frac{2C}{P}$$

(2-4.17)

2-3. Solution for parametric substitution

Example 22

Solve the differential equation

$$y = \left(\frac{dy}{dx}\right)^5 + \left(\frac{dy}{dx}\right)^3 + \frac{dy}{dx} + 5$$

(2-5.1)

Solution

Equation (2-5.1) is readily solvable for y.

To obtain the parametric representation in x, we make the following **parametric substitution**

$$\frac{dy}{dx} = P$$

(2-5.2)

i.e. $\qquad dx = \frac{dy}{P}$

(2-5.3)

Substitute from (2-5.2) and (2-5.3) into (2-5.1), we get

$$y = P^5 + P^3 + P + 5$$

(2-5.4)

$$dx = \frac{dy}{P} = \frac{d(P^5 + P^3 + P + 5)}{P}$$
$$= \frac{(5P^4 + 3P^2 + 1)}{P}dP \qquad (2\text{-}5.5)$$

Note the parametric substitution dP, which rendered equation (2-5.5) easier to integrate than without the substitution.

Integrating, we get

$$x = \int \frac{(5P^4 + 3P^2 + 1)}{P}dP$$
$$= \int (5P^3 + 3P + P^{-1})dP \qquad (2\text{-}5.6)$$
$$= \frac{5}{4}P^4 + \frac{3}{2}P^2 + \ln P + C$$

Differentiate equation (2-5.4), to get

$$\frac{dy}{dx} = 5P^4 + 3P^2 + 1 \qquad (2\text{-}5.7)$$

Summary of solution

The parametric equations of general solution are given by equations (2-5.4) and (2-5.6) as follows

$$y = P^5 + P^3 + P + 5 \qquad (2\text{-}5.4)$$
$$x = \frac{5}{4}P^4 + \frac{3}{2}P^2 + \ln P + C \qquad (2\text{-}5.6)$$

Example 23

Solve the differential equation

$$\frac{y}{\sqrt{1 + \left(\dfrac{dy}{dx}\right)^2}} = 1 \qquad (2\text{-}6.1)$$

Solution

Equation (2-6.1) is readily solvable for y.

To obtain the parametric representation in x, we make the following **parametric substitution**

$$\frac{dy}{dx} = \sinh t \qquad (2\text{-}6.2)$$

Substituting from (2-6.2) into (2-6.1), we get

$$y = \sqrt{1 + (\sinh t)^2}$$
$$= \cosh t \qquad (2\text{-}6.3)$$

Substituting by the differential of y in equation (2-6.2), we get

$$dx = \frac{dy}{\sinh t} = \frac{\sinh t}{\sinh t} dt = dt \qquad (2\text{-}6.4)$$

Integrating, we get

$$x = t + C \qquad (2\text{-}6.5)$$

Summary of solution

The parametric equations of general solution are given by equations (2-6.3) and (2-6.6) as follows

$$y = \cosh t \qquad (2\text{-}6.3)$$
$$x = t + C \qquad (2\text{-}6.6)$$

2-4. Solution for independent variable

Example 24

Solve the differential equation

$$2e^x \frac{dy}{dx} = y^2 + \left(\frac{dy}{dx}\right)^2 \qquad (2\text{-}7.1)$$

Solution

To obtain the parametric representation in x, we make the following **parametric substitution**

$$\frac{dy}{dx} = P, \qquad \frac{dx}{dy} = \frac{1}{P} \qquad\qquad (2\text{-}7.2)$$

Substituting from (2-7.2) into (2-7.1), we get

$$e^x = \frac{y^2 + P^2}{2P} \qquad\qquad (2\text{-}7.3)$$

Differentiating (2-7.3) with respect to y, we have

$$e^x \frac{dx}{dy} = \frac{d}{dy}\left(\frac{y^2 + P^2}{2P}\right)$$

$$= \frac{2P\left(2y + 2P\dfrac{dP}{dy}\right) - 2(y^2 + P^2)\dfrac{dP}{dy}}{4P^2} \qquad\qquad (2\text{-}7.4)$$

$$= \frac{2yP + P^2\dfrac{dP}{dy} - y^2\dfrac{dP}{dy}}{2P^2}$$

Substituting from (2-7.2) in (2-7.4), we get

$$e^x \frac{1}{P} = \frac{2yP + P^2\dfrac{dP}{dy} - y^2\dfrac{dP}{dy}}{2P^2}$$

i.e.,
$$e^x = \frac{1}{2P}\left(2yP + P^2\frac{dP}{dy} - y^2\frac{dP}{dy}\right) \qquad\qquad (2\text{-}7.5)$$

Equating (2-7.3) by (2-7.5), we get

$$\frac{y^2 + P^2}{2P} = \frac{1}{2P}\left(2yP + P^2\frac{dP}{dy} - y^2\frac{dP}{dy}\right)$$

$$y^2 + P^2 = 2yP + (P^2 - y^2)\frac{dP}{dy} \qquad\qquad (2\text{-}7.6)$$

Arranging, we get

59

$$\left(P^2 - y^2\right)\frac{dP}{dy} = y^2 - 2yP + P^2$$

$$(2\text{-}7.7)$$

$$= \left(P - y\right)^2$$

i.e.,
$$\frac{dy}{dP} = \frac{\left(P + y\right)}{\left(P - y\right)} = \frac{\left(1 + \frac{y}{P}\right)}{\left(1 - \frac{y}{P}\right)}$$

$$(2\text{-}7.8)$$

Substituting by

$$s = \frac{y}{P}$$
$$y = sP$$
$$\frac{dy}{dP} = P\frac{ds}{dP} + s$$

$$(2\text{-}7.9)$$

Equation (2-7.9) becomes

$$P\frac{ds}{dP} + s = \frac{\left(1 + s\right)}{\left(1 - s\right)}$$

$$P\frac{ds}{dP} = \frac{1 + s^2}{\left(1 - s\right)}$$

$$(2\text{-}7.10)$$

Putting (2-7.10) in integrable form, we get

$$\frac{\left(1 - s\right)ds}{1 + s^2} = \frac{dP}{P}$$

$$(2\text{-}7.11)$$

Integrating both sides, we get

$$\frac{ds}{1 + s^2} - \frac{sds}{1 + s^2} = \frac{dP}{P}$$

$$\ln P = \tan^{-1}s - \frac{1}{2}\ln\left(1 + s^2\right) + C$$

$$(2\text{-}7.12)$$

Substituting from (2-7.9) by s = y/P, we get

$$\ln P = \tan^{-1}\frac{y}{P} - \frac{1}{2}\ln\left(1 + \left(\frac{y}{P}\right)^2\right) + C$$

$$= \tan^{-1}\frac{y}{P} - \frac{1}{2}\ln\left(P^2 + y^2\right) + \ln P + C$$

$$(2\text{-}7.13)$$

This is reduced to

60

$$\ln(P^2 + y^2) = 2\tan^{-1}\frac{y}{P} + C \tag{2-7.13}$$

From (2-7.3), we get

$$e^x = \frac{y^2 + P^2}{2P}$$
$$x = \ln(y^2 + P^2) - \ln(2P) \tag{2-7.14}$$

Therefore, (2-7.13) becomes

$$x = 2\tan^{-1}\frac{y}{P} + C - \ln(2P) \tag{2-7.13}$$

$$\ln(2Pe^x) = 2\tan^{-1}\frac{y}{P} + C \tag{2-7.13}$$

2-4.1. Absent dependent variable in the o.d.e.

Example 25

Solve the differential equation

$$x = \left(\frac{dy}{dx}\right)^2 - \frac{dy}{dx} - 1 \tag{2-8.1}$$

Solution

To obtain the parametric representation in x, we make the following **parametric substitution**

$$\frac{dy}{dx} = P, \qquad dy = Pdx \tag{2-8.2}$$

Substituting from (2-8.2) into (2-8.1), we get

$$x = P^2 - P - 1 \tag{2-8.3}$$

And

$$dy = Pdx = Pd(P^2 - P - 1)$$
$$= P(2P - 1)dP \tag{2-8.4}$$

61

Integrating both sides, we get

$$y = \frac{2}{3}P^3 - \frac{1}{2}P^2 + C \qquad (2\text{-}8.5)$$

Summary of solution

The parametric equations of general solution are given by equations (2-8.3) and (2-8.5) as follows

$$x = P^2 - P - 1 \qquad (2\text{-}8.3)$$

$$y = \frac{2}{3}P^3 - \frac{1}{2}P^2 + C \qquad (2\text{-}8.5)$$

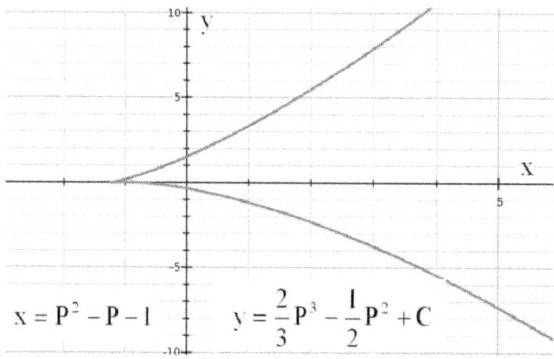

$$x = P^2 - P - 1 \qquad y = \frac{2}{3}P^3 - \frac{1}{2}P^2 + C$$

Example 26

Solve the differential equation

$$x\sqrt{1 + \left(\frac{dy}{dx}\right)^2} = \frac{dy}{dx} \qquad (2\text{-}9.1)$$

Solution

To obtain the parametric representation in x, we make the following **parametric substitution**

$$\frac{dy}{dx} = \tan t, \qquad dy = Pdx, \qquad -\frac{\pi}{2} < t < \frac{\pi}{2} \qquad (2\text{-}9.2)$$

62

Substituting from (2-9.2) into (2-9.1), we get

$$x\sqrt{1+\tan^2 t} = \tan t$$

$$x = \frac{\tan t}{\sqrt{1+\tan^2 t}} \tag{2-9.3}$$

$$x = \sin t$$

Therefore,

$$dy = \tan t\, d(\sin t)$$

$$= \tan t \cos t\, dt \tag{2-9.4}$$

$$= \sin t\, dt$$

Integrating both sides, we get

$$y = -\cos t + C \tag{2-9.5}$$

Summary of solution

The parametric equations of general solution are given by equations (2-9.3) and (2-9.5) as follows

$$x = \sin t \tag{2-9.3}$$
$$y = -\cos t + C \tag{2-9.5}$$

Arranging and eliminating t, we get

$$\sin^2 t = x^2$$

$$+ \quad \cos^2 t = (C-y)^2 \tag{2-9.6}$$

Add _____

$$1 = x^2 + (C-y)^2$$

which is a **family of circles** of radius 1 and origin (0,C).

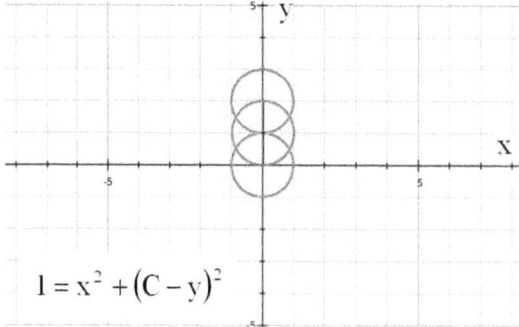

$$1 = x^2 + (C - y)^2$$

2-5. Clairaut's equation's singular and non-singular solutions

Example 27

Solve the differential equation

$$y = x\frac{dy}{dx} + f\left(\frac{dy}{dx}\right)$$

or   (2-10.1)

$$y = xy' + f(y')$$

Where f(y´) is unknown function in the first derivative of y.

Solution

Substitute by

$$\frac{dy}{dx} = P(x) \qquad (2\text{-}10.2)$$

Equation (2-10.1) becomes

$$y = xP + f(P) \qquad (2\text{-}10.3)$$

Differentiate with respect to x using the substitution in (2-10.2)

$$\frac{dy}{dx} = x\frac{dP}{dx} + P + \frac{df(P)}{dP}\frac{dP}{dx}$$
$$P = x\frac{dP}{dx} + P + \frac{df(P)}{dP}\frac{dP}{dx}$$

i.e.,

$$\left(x + \frac{df(P)}{dP}\right)\frac{dP}{dx} = 0 \qquad\qquad (2\text{-}10.4)$$

Therefore, we have the two equations

$$\frac{dP}{dx} = 0 \qquad\qquad (2\text{-}10.5a)$$

$$x + \frac{df(P)}{dP} = 0 \qquad\qquad (2\text{-}10.5b)$$

Equations (2-10.5) give the two solutions

$$P(x) = C \qquad\qquad (2\text{-}10.6a)$$
$$P(x) = g(x) \qquad\qquad (2\text{-}10.6b)$$

Where, g(x) is unknown.

Substituting by each of the two solutions of P(x) from (2-10.6) into (2-10.3), we get

$$y = xC + f(C) \qquad\qquad (2\text{-}10.7a)$$
$$y = xg(x) + f(g(x)) \qquad\qquad (2\text{-}10.7b)$$

Equation (2-10.7a) is a family of lines and comprises the general solution of **Clairaut's equation** (2-10.1).

Equation (2-10.7b) leads to **singular solution** as follows.

Substitute by y and its derivatives from (2-10.7b) into (2-10.3), we get

$$xg(x) + f(g(x)) = xg(x) + f(g(x))$$

Thus, the solution in (2-10.7b) is singular.

Example 28

Solve the differential equation

$$y = x\frac{dy}{dx} + \frac{a\dfrac{dy}{dx}}{\sqrt{1 + \left(\dfrac{dy}{dx}\right)^2}}$$

(2-11.1)

Solution

The general solution of Clairaut's equation (2-11.1) is obtained by the substitution from (2-10.6a) by

$$\frac{dy}{dx} = C$$

(2-11.2)

The general solution of equation (2-11.1) is

$$y = xC + \frac{aC}{\sqrt{1 + C^2}}$$

(2-11.3)

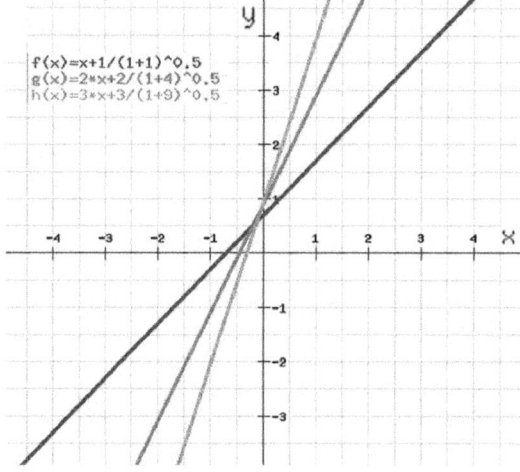

The **singular solution** of (2-11.1) is obtained by differentiating (2-11.3) with **respect to C** to get

$$\frac{dy}{dC} = x + \frac{a}{\left(1 + C^2\right)^{\frac{3}{2}}} = 0$$

(2-11.4)

Therefore,

$$x = -\frac{a}{\left(1+C^2\right)^{\frac{3}{2}}}$$ (2-11.5)

Substituting by x, from equation (2-11.5) into equation (2-11.3), we get

$$y = \frac{aC^3}{\left(1+C^2\right)^{\frac{3}{2}}}$$ (2-11.6)

Which is the singular solution and **equation of the envelope**.

Eliminating C between (2-11.5) and (2-11.6), we get

$$x^{\frac{2}{3}} + y^{\frac{2}{3}} = a^{\frac{2}{3}}$$ (2-11.7)

Which is the left branch of **astroid** set of curves.

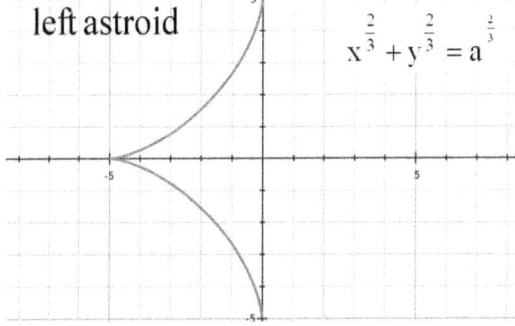

left astroid $x^{\frac{2}{3}} + y^{\frac{2}{3}} = a^{\frac{2}{3}}$

2-6. Lagrange's equation

Example 29

Solve the differential equation

$$y = xg\left(\frac{dy}{dx}\right) + f\left(\frac{dy}{dx}\right)$$

or

$$y = xg(y') + f(y')$$

(2-12.1)

Where f(y') and g(y') are unknown functions in the first derivative of y.

Solution

The Lagrange's Equation (2-12.2) is a general form of Clairaut equation, (2-10.1). (where, xg(y') in Lagrange's replaced xy', in Clairaut's).

Substitute by

$$\frac{dy}{dx} = P(x)$$

(2-12.2)

Equation (2-12.1) becomes

$$y = xg(P) + f(P)$$

(2-12.3)

Differentiate with respect to x using the substitution in (2-12.2)

$$\frac{dy}{dx} = x\frac{dg}{dP}\frac{dP}{dx} + g(x) + \frac{df}{dP}\frac{dP}{dx}$$

$$P = g(P) + \left(x\frac{dg}{dP} + \frac{df}{dP}\right)\frac{dP}{dx}$$

(2-12.4)

i.e.,

$$P - g(P) = \left(x\frac{dg}{dP} + \frac{df}{dP}\right)\frac{dP}{dx}$$

(2-12.5)

Similar to the solution of Clairaut's equation, equations (2-10.7), we have general and singular solutions as follows.

One solution is obtained from equations (2-12.5) is as follows

$$\frac{dP}{dx} = 0$$

(2-12.6a)

$$P - g(P) = 0$$

(2-12.6b)

Equations (2-12.6) give a solution for the following values of P and g:

$$P(x) = C$$

(2-12.7a)

$$g(C) = C$$

(2-12.7b)

Substituting from (2-12.7) into (2-12.3), we get

$$y = xg(C) + f(C) \tag{2-12.8}$$

Equation (2-12.8) is a family of lines.

The general solution of **Lagrange's equation** (2-12.1) is obtained by writing equation (2-12.5) in the form:

$$P - g(P) = \left(x\frac{dg}{dP} + \frac{df}{dP} \right)\frac{dP}{dx}$$

$$\frac{dx}{dP} = \left(x\frac{\dfrac{dg}{dP}}{(P - g(P))} + \frac{\dfrac{df}{dP}}{(P - g(P))} \right) \tag{2-12.9}$$

$$dx = \left(x\frac{dg}{(P - g(P))} + \frac{df}{(P - g(P))} \right)$$

Integration gives x as function on P.

Example 30

Solve the differential equation

$$y = x\left(\frac{dy}{dx}\right)^2 + \left(\frac{dy}{dx}\right)^2 \tag{2-13.1}$$

Solution

Substitute by

$$\frac{dy}{dx} = P(x) \tag{2-13.2}$$

Equation (2-13.1) is written as

$$y = xP^2 + P^2 \tag{2-13.3}$$

Differentiate with respect to x using the substitution in (2-13.2)

$$P = P^2 + 2P(x + 1)\frac{dP}{dx} \tag{2-13.4}$$

One of the two solutions is obtained by the two equations

$$\frac{dP}{dx} = 0 \qquad (2\text{-}13.5a)$$

$$P - P^2 = 0 \qquad (2\text{-}13.5b)$$

Or, $\qquad$ $P = 1$ $\qquad\qquad\qquad\qquad\qquad\qquad\qquad\qquad$ (2-13.5c)

Substituting from (2-13.5) into (2-13.3), noting that P = 1 (in this particular solution) we get

$$y = x + 1 \qquad (2\text{-}13.6)$$

The general solution is obtained similar to obtaining equation (2-12.8), by writing equation (2-13.4) in the form

$$\frac{dx}{dP} = \frac{2(x+1)}{1-P} \qquad (2\text{-}13.7)$$

Integrating, we get

$$\frac{dx}{(x+1)} = -\frac{2dP}{P-1}$$
$$\ln(x+1) = -2\ln(P-1) + 2\ln C \qquad (2\text{-}13.8)$$
$$x = -1 + \frac{C^2}{(P-1)^2}$$

Eliminating P between (2-13.3) and (2-13.8), we get the **general solution**:

$$y = \left(\sqrt{x+1} + C\right)^2 \qquad (2\text{-}13.9)$$

Since the two solutions obtained in (2-13.5) imply that $P = P^2 =$ constant, which implies that y = 0, equation (2-13.3), which cannot be obtained from equation (2-13.9), therefore, the singular solution is
$$y = 0$$

Also, since equation (2-13.6) in obtainable from the general solution (2-13.9), therefore, y=x+1 is a **particular solution**, not singular.

2-7. Differential equation of orthogonal curves

Two sets of curves are orthogonal if the angles between the tangents of two sets equal 90 degrees. This is expressed mathematically as follows:

$$\varphi\left(x, y, c, \frac{dy}{dx}\right)$$

$$\psi\left(x, y, c, -\frac{1}{\frac{dy}{dx}}\right) \qquad (2\text{-}14.1)$$

The set of curves ψ is orthogonal on the set φ, because

$$\left(\frac{dy}{dx}\right)_{\varphi}\left(-\frac{1}{\frac{dy}{dx}}\right)_{\psi} = -1$$

Example 31

The **velocity of flow** of fluid is given by velocity vector φ and its **velocity potentials** by vector ψ, equation (2-14.1) such that

$$\frac{\partial\psi}{\partial x} = \varphi_x$$

$$\frac{\partial\psi}{\partial y} = \varphi_y \qquad (2\text{-}14.2)$$

Equation (2-14.2) implies that the projections of velocities φ_x and φ_y are generated by the change in potential along the x and y directions, respectively.

The angle between the two vectors φ and ψ is given by

$$\tan\theta = \frac{\varphi_y}{\varphi_x} = \frac{\dfrac{\partial\psi}{\partial y}}{\dfrac{\partial\psi}{\partial x}} \qquad (2\text{-}14.3)$$

The **complex function** that describes the orthogonal relationship between real function vector functions $\varphi(x,y)$ and $\psi(x,y)$ can be presented as follows

$$\Psi(z) = \psi(x, y) + i\varphi(x, y) \qquad (2\text{-}14.4)$$

71

Where the xy-plane of coordinates is described by

$$z = x + iy \qquad (2\text{-}14.5)$$

Differentiating the equations (2-14.4) with respect to x and then to y, we get

$$\frac{\partial \Psi(z)}{\partial z}\frac{\partial z}{\partial x} = \frac{\partial \psi(x,y)}{\partial x} + i\frac{\partial \varphi(x,y)}{\partial x} \qquad (2\text{-}14.6a)$$

$$\frac{\partial \Psi(z)}{\partial z}\frac{\partial z}{\partial y} = \frac{\partial \psi(x,y)}{\partial y} + i\frac{\partial \varphi(x,y)}{\partial y} \qquad (2\text{-}14.6b)$$

$$\frac{\partial z}{\partial x} = 1, \qquad \frac{\partial z}{\partial y} = i \qquad (2\text{-}14.7)$$

Substituting from (2-14.7) in (2-14.6), we get

$$\frac{\partial \Psi(z)}{\partial z} = \frac{\partial \psi(x,y)}{\partial x} + i\frac{\partial \varphi(x,y)}{\partial x} \qquad (2\text{-}14.8a)$$

$$i\frac{\partial \Psi(z)}{\partial z} = \frac{\partial \psi(z)}{\partial y} + i\frac{\partial \varphi(z)}{\partial y} \qquad (2\text{-}14.8b)$$

Multiply the last equation by –I, we get

$$\frac{\partial \Psi(z)}{\partial z} = -i\frac{\partial \psi(x,y)}{\partial y} + \frac{\partial \varphi(x,y)}{\partial y} \qquad (2\text{-}14.8c)$$

Equate the real and imaginary parts between equations (2-14.8a) and (2-14.8c), we get

$$\frac{\partial \psi(x,y)}{\partial y} = -\frac{\partial \varphi(x,y)}{\partial x} \qquad (2\text{-}14.9a)$$

$$\frac{\partial \psi(x,y)}{\partial x} = +\frac{\partial \varphi(x,y)}{\partial y} \qquad (2\text{-}14.9b)$$

Dividing the above two equations by member, we get

$$\tan\theta_1 = \frac{\dfrac{\partial \psi(x,y)}{\partial y}}{-\dfrac{\partial \varphi(x,y)}{\partial x}} = -1 \qquad (2\text{-}14.10a)$$

$$\tan\theta_2 = \frac{\dfrac{\partial\psi(x,y)}{\partial x}}{\dfrac{\partial\phi(x,y)}{\partial y}} = 1 \tag{2-14.10b}$$

Equation (2-14.10a) represents a set of **orthogonal curves**.
Equation (2-14.10b) represents a set of **parallel curves**.

Example 32

Find the orthogonal projections of the family of parabolas

$$y = Cx^2 \tag{2-15.1}$$

Solution

Differentiate equation (2-15.1)

$$y' = 2Cx \tag{2-15.2}$$

Divide the two equations above to eliminate the constant C

$$\frac{y'}{y} = \frac{2}{x} \tag{2-15.3}$$

The **orthogonal trajectories** are obtained by replacing y' by -1/y' in the above equations:

$$-\frac{1}{y'y} = \frac{2}{x} \tag{2-15.4}$$

Integrating, we get

$$\frac{dy}{dx} = -\frac{x}{2y}$$
$$\int 2y\,dy = -\int x\,dx \tag{2-15.5}$$
$$y^2 + \frac{x^2}{2} = C$$

2-8. Differential equation of isogonal curves

Two sets of curves ψ and φ are **isogonal** if the angles θ between the tangents of two sets is constant.

$$\tan\theta = k \tag{2-16.1}$$

Let one set of the two curves has the angle α given by

$$\tan\alpha = \frac{d\varphi}{dx} \tag{2-16.2}$$

Let one set of the two curves has the angle β given by

$$\tan\beta = \frac{d\psi}{dx} \tag{2-16.3}$$

Where

$$\alpha = \beta - \theta \tag{2-16.4}$$

Therefore,

$$\begin{aligned}
\tan\alpha &= \tan(\beta - \theta) \\
&= \frac{\tan\beta - \tan\theta}{1 + \tan\beta\tan\theta} \\
&= \frac{\tan\beta - k}{1 + k\tan\beta}
\end{aligned} \tag{2-16.5}$$

Substituting by the differential derivatives of slopes, we get

$$\frac{d\varphi}{dx} = \frac{\dfrac{d\psi}{dx} - k}{k\dfrac{d\psi}{dx} + 1} \tag{2-16.6}$$

Example 33

Find the isogonal projections of the family of parabolas

$$y = Cx^2 \tag{2-17.1}$$

Solution

Differentiate equation (2-17.1)

$$y' = 2Cx \qquad (2\text{-}17.2)$$

Divide the two equations above to eliminate the constant C

$$y' = \frac{2y}{x} \qquad (2\text{-}17.3)$$

The **isogonal trajectories** making angle θ, (tanθ = k), are obtained from equation (2-16.6), as follows

$$\frac{dy}{dx} = \frac{\dfrac{d\psi}{dx} - k}{k\dfrac{d\psi}{dx} + 1} \qquad (2\text{-}17.4)$$

$$= \frac{2y}{x}$$

Solving for the $\dfrac{d\psi}{dx}$, we get

$$\frac{d\psi}{dx} - k = \frac{2y}{x}\left(k\frac{d\psi}{dx} + 1\right) \qquad (2\text{-}17.5a)$$

Or,

$$\frac{d\psi}{dx} = \frac{1}{k}\frac{\dfrac{k}{2} + \dfrac{y}{x}}{\left(\dfrac{1}{2k} - \dfrac{y}{x}\right)} \qquad (2\text{-}17.5b)$$

Integrating the homogeneous equation, (ψ = y) we get

$$\frac{1}{2}\ln\left(k\left(2y^2 + x^2\right) + xy\right) = \frac{3\tan^{-1}\left(\dfrac{4ky + x}{x\left(\sqrt{8k^2 - 1}\right)}\right)}{\sqrt{8k^2 - 1}} + C \qquad (2\text{-}17.5b)$$

Which defines the set of curves that make angle θ = tan⁻¹k, equation (2-16.1), with the set y = Cx², equation (2-17.1).

2-9. Second-Order o.d.e. reducible to First-Order

2-9.1. Differential equations devoid of the dependent variable

Example 34

Solve the second-order o.d.e. by reducing it to first-order

$$\frac{d^2y}{dx^2} = \frac{1}{a}\sqrt{1+\left(\frac{dy}{dx}\right)^2} \qquad (2\text{-}18.1)$$

Solution

Equation (2-18.1) is devoid of the explicit variable y. Therefore, it can be integrated with respect to **the first derivative as an independent variable**, then the result of integration is integrated again with respect to the **initial independent variable, x**.

Substitute by

$$P = \frac{dy}{dx} \qquad (2\text{-}18.2)$$

Then $\qquad \frac{dP}{dx} = \frac{d^2y}{dx^2} \qquad (2\text{-}18.3)$

Substitute from (2-18.2) and (2-18.3) in (2-18.1), we get the **first-order o.d.e. in P**, as follows:

$$\frac{dP}{dx} = \frac{1}{a}\sqrt{1+P^2} \qquad (2\text{-}18.4)$$

Separating variables and integrating, we get

$$\int \frac{dP}{\sqrt{1+P^2}} = \frac{1}{a}\int dx$$
$$\ln\left(P+\sqrt{1+P^2}\right) = \frac{x}{a}+C \qquad (2\text{-}18.5)$$

Arranging, we get

$$\sqrt{1+P^2} = e^{\frac{x}{a}+C} - P$$
$$1+P^2 = e^{2\left(\frac{x}{a}+C\right)} + P^2 - 2Pe^{\frac{x}{a}+C} \qquad (2\text{-}18.6)$$
$$P = \frac{1}{2}\left(e^{\left(\frac{x}{a}+C\right)} - e^{-\left(\frac{x}{a}+C\right)}\right)$$

Substituting from equation (2-18.2), and integrating, we get

$$\frac{dy}{dx} = \frac{1}{2}\left(e^{\left(\frac{x}{a}+C\right)} - e^{-\left(\frac{x}{a}+C\right)} \right)$$

$$y = \frac{1}{2}\int \left(e^{\left(\frac{x}{a}+C\right)} - e^{-\left(\frac{x}{a}+C\right)} \right) dx \tag{2-18.7}$$

$$= \frac{a}{2}\left(e^{\left(\frac{x}{a}+C\right)} + e^{-\left(\frac{x}{a}+C\right)} \right) + C'$$

This is the general solution of equation (2-18.1).

Particular Solution

Give the boundary conditions

x=0,
y(0) = a, and
y'(0) = 0 $\qquad$ (2-18.8)

Substituting by the boundary conditions in equation (2-18.6), we get

$$P = \frac{1}{2}\left(e^{\left(\frac{x}{a}+C\right)} - e^{-\left(\frac{x}{a}+C\right)} \right)$$

$$0 = \frac{1}{2}\left(e^{(0+C)} - e^{-(0+C)} \right)$$

$$e^{(0+C)} = e^{-(0+C)} \tag{2-18.9}$$

$$C = -C$$

$$C = 0$$

Substituting by the boundary conditions in equation (2-18.7), we get

$$y = \frac{a}{2}\left(e^{\left(\frac{x}{a}+C\right)} + e^{-\left(\frac{x}{a}+C\right)} \right) + C'$$

$$a = \frac{a}{2}\left(e^{0} + e^{-0} \right) + C' \tag{2-18.10}$$

$$a = a + C'$$

$$C' = 0$$

77

Thus, the general solution, (2-18.7), becomes

$$y = \frac{a}{2}\left(e^{\left(\frac{x}{a}\right)} + e^{-\left(\frac{x}{a}\right)}\right)$$
$$= a\cosh\frac{x}{a}$$

(2-18.11)

2-9.2. Differential equations devoid of the independent variable

Example 35

Solve the second-order o.d.e. by reducing it to first-order

$$3\frac{d^2y}{dx^2} = y^{-\frac{5}{3}}$$

(2-19.1)

Solution

Equation (2-19.1) is devoid of the explicit variable x. Therefore, it can be integrated with respect to **the first derivative as an independent variable**, then the result of integration is integrated again with respect to the **initial dependent variable, y**.

Substitute by

$$P(y) = \frac{dy}{dx}$$

(2-19.2)

Where, P(y) is considered function in y, which in turn is function in x.

Therefore,

$$\frac{d^2y}{dx^2} = \frac{dP}{dy}\frac{dy}{dx}$$
$$= P\frac{dP}{dy}$$

(2-19.3)

Substitute from (2-19.2) and (2-19.3) in (2-19.1), we get the **first-order o.d.e. in P**, as follows:

78

$$3P\frac{dP}{dy} = y^{-\frac{5}{3}}$$

(2-19.4)

Separating variables and integrating, we get

$$3\int PdP = \int y^{-\frac{5}{3}}dy$$

$$\frac{3}{2}P^2 = -\frac{3}{2}y^{-\frac{2}{3}} + C$$

(2-19.5)

$$P = \sqrt{C - y^{-\frac{2}{3}}}$$

The second integration process requires substituting from (2-19.2) into (2-19.5), separating variables, and integrating, as follows

$$\frac{dy}{dx} = \sqrt{C - y^{-\frac{2}{3}}}$$

$$\pm \frac{dy}{\sqrt{C - y^{-\frac{2}{3}}}} = dx$$

(2-19.6)

Or,

$$\pm \int \frac{y^{\frac{1}{3}}dy}{\sqrt{y^{\frac{2}{3}}C - 1}} = \int dx = x + C$$

(2-19.7)

To determine the remaining integral, substitute by

$$y^{\frac{2}{3}}C - 1 = t^2$$

(2-19.8a)

And

$$y^{\frac{2}{3}} = \frac{t^2 + 1}{C}$$

(2-19.8b)

This yields the differentials

$$\frac{2}{3}y^{-\frac{1}{3}}Cdy = 2tdt$$

(2-19.9)

Making the substitutions of t and its differential in (2-19.7), we get

$$x + C = \pm \int \frac{3y^{\frac{2}{3}} t\, dt}{Ct}$$

$$= \pm \frac{3}{C^2} \int \left(t^2 + 1\right) dt \qquad\qquad (2\text{-}19.10)$$

$$= \pm \frac{3}{C^2} \left(\frac{t^3}{3} + t\right) + C'$$

C' is a constant.

Reverting from t to y, by substituting from equation (2-19.8a), we get

$$x + C = \pm \frac{1}{C^2} \left(y^{\frac{2}{3}} C + 2\right) \sqrt{y^{\frac{2}{3}} C - 1} + C' \qquad\qquad (2\text{-}19.11)$$

2-10. Exercises on first-order of higher degree o.d.e.

Solve the following o.d.e.

2-1. $\quad xy\left(\dfrac{dy}{dx}\right)^2 - (x^2 + y^2)\dfrac{dy}{dx} + xy = 0$ [Ans: $(y + cx)(y^2 - x^2 + C) = 0$]

2-2. $\quad \left(\dfrac{dy}{dx}\right)^2 = 9y^4$ $\qquad\qquad\qquad$ [Ans: $y^2 = \dfrac{1}{(3x + C)^2}$]

2-3. $\quad x^2 + \left(\dfrac{dy}{dx}\right)^2 = 1$ $\qquad\qquad\qquad$ Put y' = cost, x = sint

$$\text{Ans:} \left[y = \frac{1}{2}t + \frac{1}{4}\sin 2t + C \right]$$

2-4. $\quad x = \left(\dfrac{dy}{dx}\right)^2 - \dfrac{dy}{dx} + 2$ $\quad \left[\text{Ans}:\ x = P^3 - P + 2,\quad y = \dfrac{3}{4}P^4 - \dfrac{1}{2}P^2 + C \right]$

2-5. $\quad y = \left(\dfrac{dy}{dx}\right)^4 - \left(\dfrac{dy}{dx}\right)^3 - 2$ $\quad \left[\text{Ans}:\ x = \dfrac{4}{3}P^3 - \dfrac{3}{2}P^2 + C,\quad y = P^4 - P^3 - 2C \right]$

2-6. $\quad y = 2x\dfrac{dy}{dx} - \left(\dfrac{dy}{dx}\right)^2$ $\quad \left[\text{Ans}:\ x = \dfrac{C}{P^4} + \dfrac{2}{3}P,\quad y = 2xP - P^2 \right]$

2-7. $\quad y = x\dfrac{dy}{dx} + \left(\dfrac{dy}{dx}\right)^2$, $\quad$ for $\quad$ x = 2, $\quad$ y = -1 $\quad \left[\text{Ans}:\ y = -x + 1,\quad y = -\dfrac{x^2}{4} \right]$

2-8. $\quad y = x\dfrac{dy}{dx} + \left(\dfrac{dy}{dx}\right)^2$, $\quad$ for $\quad$ x = 1, $\quad$ y = -1 $\quad$ [no real solution exits]

2-9. $\quad y = x^2 + 2x\dfrac{dy}{dx} + \dfrac{1}{2}\left(\dfrac{dy}{dx}\right)^2$ $\quad \left[\begin{array}{l} \text{Ans}:\ \text{general solution}\ \ y = cx + \dfrac{1}{2}(c^2 - x^2) \\ \quad\quad \text{singular solution}\ \ y = -x^2 \end{array} \right]$

2-10. $\quad y\left(1+\left(\dfrac{dy}{dx}\right)^2\right)=a$

$$\left[\begin{array}{l}\text{Hint : put } y'=\cot t \\[4pt] \text{Ans: } x-c=\dfrac{a}{2}(2t-\sin 2t), \qquad y-c=\dfrac{a}{2}(t-\cos 2t) \\[4pt] \text{a family of cycloids,} \qquad \text{a singular solution is } y=a \end{array}\right]$$

2-11. $\quad \left(\dfrac{dy}{dx}\right)^2+(x+a)\dfrac{dy}{dx}-y=0, \qquad$ where a is constant.

$$\left[\begin{array}{l}\text{Ans: general solution } y=c(x+a)+c^2 \\[6pt] \text{singular solution } y=-\dfrac{1}{4}(x+a)^2 \end{array}\right]$$

2-12. $\quad y=\left(\dfrac{dy}{dx}\right)^2+2x\dfrac{dy}{dx}$ $\qquad \left[\text{Ans: } x=\dfrac{C}{3P^2}-\dfrac{2}{3}P,\ y=\dfrac{2C-P^3}{3P}\right]$

2-13. $\quad y=x\left(\dfrac{dy}{dx}\right)^2+\left(\dfrac{dy}{dx}\right)^2$ $\qquad \left[\begin{array}{l}\text{Ans: } y=(\sqrt{x+1}+C)^2 \\ \text{singular solution } y=0.\end{array}\right]$

2-14. $\quad y=x\left(1+\dfrac{dy}{dx}\right)+\left(\dfrac{dy}{dx}\right)^2$ $\qquad \left[\begin{array}{l}\text{Ans: } x=Ce^{-P}-2P+2, \\[4pt] \quad y=C(P+1)e^{-P}-P^2+2\end{array}\right]$

2-15. $\quad y=2x\dfrac{dy}{dx}+y\left(\dfrac{dy}{dx}\right)^2$ $\qquad \left[\text{Ans: } 2Cx=4C^2-y^2\right]$

2-16. $\quad y=(x+1)\dfrac{dy}{dx}-\left(\dfrac{dy}{dx}\right)^2$ $\qquad \left[\begin{array}{l}\text{Ans: } y=Cx+C-C^2 \\ \text{singular solution } 4y=(x+1)^2.\end{array}\right]$

2-17. $\quad y=x\dfrac{dy}{dx}+\sqrt{1-\left(\dfrac{dy}{dx}\right)^2}$ $\qquad \left[\text{Ans: } y=Cx+C\right]$

2-18. $y = x\dfrac{dy}{dx} + \dfrac{dy}{dx}$

$$\left[\begin{array}{l} \text{Ans}: \ y = Cx + C - C^2 \\ \text{singular solution } 4y = (x+1)^2. \end{array}\right]$$

2-19. $y = x\dfrac{dy}{dx} + \dfrac{1}{\left(\dfrac{dy}{dx}\right)}$

$$\left[\begin{array}{l} \text{Ans}: \ y = Cx + \dfrac{1}{C} \\ \text{singular solution } y^2 = 4x \end{array}\right]$$

2-20. $y = x\dfrac{dy}{dx} - \dfrac{1}{\left(\dfrac{dy}{dx}\right)^2}$

$$\left[\begin{array}{l} \text{Ans}: \ y = Cx - \dfrac{1}{C^2} \\ \text{singular solution } y^3 = -\dfrac{27}{4}x^2 \end{array}\right]$$

2-21. Given the set of curves $xy = C$. Find the orthogonal set of curves

$$\left[\text{Ans}: \ \text{hyperbolas} \quad x^2 - y^2 = C \right]$$

2-22. Given the set of curves $x^2 + y^2 = 2\,ax$. Find the orthogonal set of curves. Use polar coordinates.

$$\left[\text{Ans}: \quad x^2 + y^2 - 2cy = 0 \right]$$

2-23. Given the set of curves $y^2 + 2ax = a^2$. Find the orthogonal set of curves. Use polar coordinates.

$$\left[\text{Ans}: \quad y^2 = 2Cx + C^2. \right]$$

2-24. Find whether on not the following o.d.e. has singular solution

$y = 5x\dfrac{dy}{dx} - \left(\dfrac{dy}{dx}\right)^2$ $\qquad \left[\text{Ans}: \ \text{No singular solution} \right]$

2-25. Find whether on not the following o.d.e. has singular solution

$\dfrac{dy}{dx} = (x - 5y)^{\frac{3}{2}} + 2$ $\qquad \left[\text{Ans}: \ \text{No singular solution} \right]$

2-26. Given the set of curves $y^2 = Cx^3$. Find the orthogonal set of curves.

$$\left[\text{Ans:} \qquad x^2 + \frac{3}{2}y^2 = c^2\right]$$

2-27. Given the set of curves $y = a\,x^n$. Find the orthogonal set of curves.

$$\left[\text{Ans:} \qquad x^2 + ny^2 = C\right]$$

2-28. Given the set of curves $y^2 = 2P(x - \alpha)$. Find the orthogonal set of curves.

Where, α is constant. $\quad \left[\text{Ans:} \qquad y = Ce^{-\frac{x}{P}}\right]$

2-29. Given the set of curves $x^2 = 2\alpha(y - x\sqrt{3})$.
Find the isogonal set of curves making angle $60°$ with the given set. .

$$\left[\text{Ans:} \qquad y^2 = C(x - y\sqrt{3})\right]$$

2-30. Given the set of curves $y^2 = 4Cx$.
Find the isogonal set of curves making angle $45°$ with the given set.

$$\left[\text{Ans:} \ y^2 - xy + 2x^2 = Ce^{\frac{6}{\sqrt{7}}\tan^{-1}\left(\frac{2y-x}{x\sqrt{7}}\right)}\right]$$

Solve the following second-order o.d.e. by reduction to first-order o.d.e.

2-31. $x\dfrac{d^2y}{dx^2} - \dfrac{dy}{dx} = x^2e^x$ $\qquad \left[\text{Ans:} \ y = e^x(x-1) + Ax^2 + B.\right]$

2-32. $x\dfrac{d^2y}{dx^2} - \left(\dfrac{dy}{dx}\right)^2 + \left(\dfrac{dy}{dx}\right)^3 = x^2e^x$ $\qquad \left[\text{Ans:} \ y + \ln y = x + B\right]$

2-33. $\dfrac{d^2y}{dx^2} + \dfrac{dy}{dx}\tan x = \sin 2x$ $\qquad \left[\text{Ans:} \ y = A + B\sin x - x - \dfrac{1}{2}\sin 2x\right]$

2-34. $\left(\dfrac{d^2y}{dx^2}\right)^2 + \left(\dfrac{dy}{dx}\right)^2 = a^2$ $\qquad \left[\text{Ans:} \ y = A - a\cos(x + B)\right]$

CHAPTER 3

FIRST-ORDER LINEAR DIFFERENTIAL EQUATIONS WITH CONSTANT COEFFICIENTS

3-1. Characteristics of linear differential equations

An ordinary differential equation of the form

$$\sum_{n=0}^{n} a_n(x)\frac{d^n y(x)}{dx^n} = f(x) \tag{3-1}$$

is **linear** in the derivatives of $y(x)$ with respect to x, when all coefficients of derivatives are only functions in x.

The linear o.d.e., in equation (3-1), is **homogeneous** is $f(x) = 0$.

The linear o.d.e., in equation (3-1), is **non-homogeneous** is $f(x) \neq 0$.

The linear o.d.e. is written in terms of **differential operator D** as follows

$$\left(\sum_{n=0}^{n} a_n(x)D^n\right) y(x) = f(x) \tag{3-2}$$

Where, the linear differential operator D is defined as

$$D = \frac{d}{dx}, \qquad D^n = \frac{d^n}{dx^n} \tag{3-3}$$

3-1.1. Properties of the linear differential operator D

1. Factorization

D-operation of **factorized variable** y is equal the factorized D-operation of variable y.

$$\left(\sum a_n(x)D^n\right)Cy(x) = C\left(\sum a_n(x)D^n\right)y(x) \tag{3-4.1}$$

The **operator polynomial** with constant coefficients $\sum a_n D^n$ can be expressed in the function form as

$$F(D)y = \sum a_n D^n y = f(x) \tag{3-4.2}$$

2. Sums of operands

D-operation of a **sum of variables** y is equal the sum of D-operation of individuals of variables y.

$$\left(\sum a_n(x)D^n\right)(y_1(x) + y_2(x)) = \left(\sum a_n(x)D^n\right)y_1(x) + \left(\sum a_n(x)D^n\right)y_2(x) \tag{3-5}$$

3. Sums of factorized operands

D-operation of a **sum of factorized variables** y is equal to the sum of factorized D-operation of individuals of variables y.

$$\left(\sum a_n(x)D^n\right)(C_1 y_1(x) + C_2 y_2(x)) = C_1\left(\sum a_n(x)D^n\right)y_1(x) + C_2\left(\sum a_n(x)D^n\right)y_2(x) \tag{3-6}$$

4. Factorized solution operand

(a) Constant multiplication: $F(D)Cy = C\,F(D)y$

If y_1 is the solution of the linear homogeneous o.d.e. (3-1), (f(x) = 0), then the **product of y_1 by an arbitrary constant** (Cy_1) is also a solution of the same homogeneous o.d.e. equation.

Since $\qquad \sum_{n=0}^{n} a_n(x)\dfrac{d^n y_1(x)}{dx^n} = 0$

Therefore,

86

$$\sum a_n(x)\frac{d^n(Cy_1(x))}{dx^n} = 0$$

$$C\sum a_n(x)\frac{d^n y_1(x)}{dx^n} = 0$$

(3-7.1)

(b) Product of D-operator polynomials: $F_1(D)F_2(D)y = F_1(D)\ [F_2(D)y]$ (3-7.2)

The product of two differential polynomials is executed in sequence.

(c) Sum of D-operator polynomials: $[F_1(D)+F_2(D)]y = F_1(D)y + F_2(D)y$ (3-7.3)

The sum of two differential polynomials is distributive.

5. Sums of solution operands

If y_1 and y_2 are solutions of the linear homogeneous o.d.e., then their **sum** $(y_1 + y_2)$ is also a solution of the same linear homogeneous o.d.e.

Since,

$$\sum a_n(x)\frac{d^n y_1(x)}{dx^n} = 0$$

$$\sum a_n(x)\frac{d^n y_2(x)}{dx^n} = 0$$

Therefore,

$$\sum a_n(x)\frac{d^n[y_1(x)+y_2(x)]}{dx^n} = 0$$

$$\sum a_n(x)\frac{d^n y_1(x)}{dx^n} + \sum a_n(x)\frac{d^n y_2(x)}{dx^n} = 0$$

(3-8)

6. Sums of factorized solution operands

If y_1 and y_2 are solutions of the linear homogeneous o.d.e., then the **sum of their factorized members** $(C_1y_1 + C_2y_2)$ is also a solution of the same linear homogeneous o.d.e.

87

Since,

$$\sum a_n(x)\frac{d^n y_1(x)}{dx^n} = 0$$

$$\sum a_n(x)\frac{d^n y_2(x)}{dx^n} = 0$$

Therefore,

$$\sum a_n(x)\frac{d^n[C_1 y_1(x) + C_2 y_2(x)]}{dx^n} = 0$$

$$C_1\left(\sum a_n(x)\frac{d^n y_1(x)}{dx^n}\right) + C_2\left(\sum a_n(x)\frac{d^n y_2(x)}{dx^n}\right) = 0$$

(3-9)

7. Sums of complex operands

If the **complex function** $(y_1 + i y_2)$ is a solution of the linear homogeneous o.d.e., then both the real y_1 and imaginary y_2 parts, each is also a solution of the same linear homogeneous o.d.e.

Since, $$\sum a_n(x)\frac{d^n[y_1(x) + i y_2(x)]}{dx^n} = 0$$ (3-10)

Therefore,

$$\left(\sum a_n(x)\frac{d^n y_1(x)}{dx^n}\right) + i\left(\sum a_n(x)\frac{d^n y_2(x)}{dx^n}\right) = 0$$

Equating the real and imaginary parts of both sides, we get

$$\sum a_n(x)\frac{d^n y_1(x)}{dx^n} = 0$$

$$\sum a_n(x)\frac{d^n y_2(x)}{dx^n} = 0$$

8. Linearly dependency of functions

Linearly dependent functions are related by homogenous equation, (3-6).
Linearly independent functions are related by non-homogenous equation.

Since in a given interval, $a \le x \le b$, we can write

$$C_1 y_1(x) + C_2 y_2(x) + C_3 y_3(x) = 0 \qquad (3\text{-}11.1)$$
$$A_1 z_1(x) + A_2 z_2(x) + A_3 z_3(x) \neq 0 \qquad (3\text{-}11.2)$$

The functions y_1, y_2, y_3 are linearly **dependent** if there exist no vanishing arbitrary constants C's such that (1-11.1) is satisfied.

The functions z_1, z_2, z_3 are linearly **independent** if there exist no vanishing arbitrary constants A's such that (1-11.2) cannot be avoided unless A's vanish.

9. Vanishing Wronskian determinant of equations linearly dependent functions

Given linearly **dependent functions** y_1, y_2, ..y_n.

Therefore

$$\sum_{i=1}^{n} C_i y_i(x) = 0 \qquad (3\text{-}12.1)$$

Differentiating the linear homogenous equation (3-12.1) (n-1) times we get (n-1) linear homogeneous o.d.e. as follows

$$D^j \left(\sum_{i=1}^{n} C_i y_i(x) \right) = 0 \qquad (3\text{-}12.2)$$

Where j varies from 1 to n-1.

Equations (3-12.1) and (3-12.2) comprise n-simultaneous linear homogeneous o.d.e., which **Wronskian determinant**, at every point x, in the interval $a \leq x \leq b$, vanishes as follows

$$W(y_i(x)) = \begin{vmatrix} y_i & y_2 & .. & y_n \\ y_i' & y_2' & .. & y_n' \\ .. & .. & .. & .. \\ y_i^{n-1} & y_2^{n-1} & .. & y_n^{n-1} \end{vmatrix} = 0 \qquad (3\text{-}12.3)$$

Since all coefficients Ci are greater than zero, from the definition of dependent functions, equation (3-11.1).

10. Non-vanishing Wronskian determinant of equations linear independent functions

Given linearly **independent functions** $y_1, y_2,..y_n$.

Therefore

$$\sum_{i=1}^{n} C_i y_i(x) \neq 0 \tag{3-13.1}$$

If y_i are solutions of the homogeneous linear equations

$$\left(\sum_{i=1}^{n} a_i(x)D^i\right) y(x) = 0 \tag{3-13.2}$$

Where $a_i(x)$ are continuous coefficients.

Equations (3-13.1) and (3-13.2) comprise n-simultaneous linear non-homogeneous o.d.e., which **Wronskian determinant**, at every point x, in the interval $a \leq x \leq b$, **does not vanish** since all coefficients Ci are not greater than zero, from the definition of independent functions, equation (3-11.2).

11. General Solution of linear combinations of solutions of homogeneous o.d.e.

According to equations (3-13.2), we could only have (n-1) differential equations of the sum combinations of all of a homogeneous o.d.e. in addition to the non-homogeneous equation (3-13.1). Thus, the number of linear **independent solutions** of a homogeneous linear o.d.e. is equal to its order.

Example 36

Given the o.d.e.

$$y''(x) - y(x) = 0 \tag{3-14.1}$$

Find the maximum number of solutions.

Solution

Substitute by $\quad P(y) = \dfrac{dy}{dx}$ $\tag{3-14.2}$

Equation (3-14.1) becomes

$$\frac{dP}{dy}\frac{dy}{dx} = y$$

$$P\frac{dP}{dy} = y \qquad (3\text{-}14.3)$$

Integrating once, we get

$$PdP = ydy \qquad (3\text{-}14.4)$$

$$\frac{1}{2}P^2 = \frac{y^2}{2}$$

i.e.,

$$P = \pm y \qquad (3\text{-}14.5)$$

Integrate twice, we get

$$\frac{dy}{y} = \pm dx$$

$$\ln y = \pm x \qquad (3\text{-}14.6)$$

$$y = e^{\pm x}$$

Therefore, for the second-order o.d.e. (3-14.1), we have the **two solutions** given in (3-14.6).

Using the property of sum of factorized solutions, equations (3-9), we get the general solution

$$y = C_1 e^x + C_2 e^{-x} \qquad (3\text{-}14.7)$$

Example 37

Given the o.d.e.

$$y'''(x) - y'(x) = 0 \qquad (3\text{-}15.1a)$$

Prove that

$$y = C_1 e^x + C_2 \cosh x + C_3 \sinh x \qquad (3\text{-}15.1b)$$

Is **not** the general solution of (3-15.1a)

Solution

Substitute by

91

$$P(y) = \frac{dy}{dx}$$

$$\frac{dP}{dy}P = \frac{d^2y}{dx^2}$$

(3-15.2)

After one integration with respect to x, and substituting from (3-15.2), equation (3-15.1a) becomes

$$\frac{dP}{dy}P = Cy$$

(3-15.3)

Where C is the first integration constant.

The second integration with respect to y gives

$$\frac{P^2}{2} = C\frac{y^2}{2}$$

(3-15.4a)

Or,

$$P = \pm Cy$$

(3-15.4b)

Substituting by P(y), from (3-15.2), we get

$$\frac{dy}{y} = \pm Cdx$$

(3-15.5)

The last integration gives

$$y = C_1 e^x + C_2 e^{-x} + C_3$$

(3-15.6)

Since the three functions in (3-15.1b) are linearly dependent, their sum comprises particular solution. Since

$$y = C_1 e^x + C_2 \left(e^x + e^x\right) + C_3 \left(e^x - e^x\right)$$

(3-15.7b)

Where the three constants could be managed to yield a vanishing sum.
In contrast, in equation (3-15.6), the sum cannot vanish for any nonvanishing constants.

3-1.2. Linear differential D-operator of some elementary functions

1- D-Operator on exponential function

$$F(D)e^{\alpha x} = e^{\alpha}F(\alpha)$$

(3-16.1)

Proof

$$F(D)e^{\alpha x} = \sum a_n D^n e^{\alpha x}$$

$$= e^{\alpha x} \sum a_n \alpha^n \qquad (3\text{-}16.2)$$

$$= e^{\alpha x} F(\alpha)$$

2- D-Operator on sine function

$$F(D^2)\sin \beta x = \sin \beta x F(-\beta^2) \qquad (3\text{-}17.1)$$

Proof

$$F(D^2)\sin \beta x = \sum a_n D^{2n} \sin \beta x$$

$$= \sin \beta x \sum a_n (-\beta^2)^n \qquad (3\text{-}17.2)$$

$$= \sin \beta x F(-\beta^2)$$

3- D-Operator on cosine function

$$F(D^2)\cos \beta x = \cos \beta x F(-\beta^2) \qquad (3\text{-}18.1)$$

Proof

$$F(D^2)\cos \beta x = \sum a_n D^{2n} \cos \beta x$$

$$= \cos \beta x \sum a_n (-\beta^2)^n \qquad (3\text{-}18.2)$$

$$= \cos \beta x F(-\beta^2)$$

4- D-Operator on product of exponential function by other general function

$$F(D)e^{\alpha x} g(x) = e^{\alpha x} F(D + \alpha)g(x) \qquad (3\text{-}19.1)$$

Proof

$$F(D)e^{\alpha x}g(x) = \sum_{k=0}^{n} a_k D^k \left(e^{\alpha x}g(x)\right)$$

$$= e^{\alpha x}\sum_{k=0}^{n} a_k \left(\alpha^k g(x) + k\alpha^{k-1}Dg(x) + \frac{k(k-1)}{2!}\alpha^{k-2}D^2 g(x) + ... + D^k g(x)\right) \qquad (3\text{-}19.2)$$

$$= e^{\alpha x}\sum_{k=0}^{n} a_k (D+\alpha)^k g(x)$$

$$= e^{\alpha x}F(D+\alpha)g(x)$$

3-2. Reduction of the order of homogeneous linear o.d.e.

The **order n** of the homogeneous linear differential equation given by

$$\left(\sum_{i=1}^{n} a_i(x)D^i\right)y(x) = 0 \qquad (3\text{-}20.1)$$

can be reduced in **order k**, without affecting the homogeneity or linearity, if k number of nontrivial **particular solutions** of (3-20.1) are known, as follows.

Given a particular solution y_1 for the homogeneous o.d.e. (3-20.1), substituting by

$$y = y_1 \int u dx \qquad (3\text{-}20.2)$$

Where u is defined as

$$\frac{dz}{dx} = u, \qquad z = \int u dx \qquad (3\text{-}20.3)$$

And

$$y = y_1 z \qquad (3\text{-}20.4)$$

Thus, equation (3-20.1) becomes

$$\left(\sum_{i=1}^{n} a_i(x)D^i\right)y_1 z = 0 \qquad (3\text{-}20.5)$$

Thus, if $z = 1$, $y_1 a_0 = 0$, i.e., $a_0 = 0$, which reduces the order of the differential by one. The same process can be repeated as many as k number of known solution.

Example 38

Knowing that $y = x$ is a particular solution of the equation

$$x\frac{d^2y}{dx^2} - x\frac{dy}{dx} + y = 0 \qquad (3\text{-}21.1)$$

Show that the order of the equation could be reduced by unity.

Solution

From equation (3-20.4), substitute by

$$y = xz \qquad (3\text{-}21.2)$$

Differentiate twice we get

$$\frac{dy}{dx} = x\frac{dz}{dx} + z$$
$$\frac{d^2y}{dx^2} = x\frac{d^2z}{d^2x} + 2\frac{dz}{dx} \qquad (3\text{-}21.3)$$

Substituting from (3-21.3) in equation (3-21.1), we get

$$x\left(x\frac{d^2z}{d^2x} + 2\frac{dz}{dx}\right) - x\left(x\frac{dz}{dx} + z\right) + xz = 0$$
$$x^2\frac{d^2z}{d^2x} - x(2-x)\frac{dz}{dx} = 0 \qquad (3\text{-}21.4)$$

Or,
$$\frac{d^2z}{d^2x} - \frac{(2-x)}{x}\frac{dz}{dx} = 0 \qquad (3\text{-}21.5)$$

Denoting

$$u = \frac{dz}{dx}, \qquad \frac{du}{dx} = \frac{d^2z}{dx^2} \qquad (3\text{-}21.6)$$

Therefore, (3-21.5) becomes

Or,
$$\frac{du}{dx} = \frac{(2-x)}{x}u \qquad (3\text{-}21.7)$$

Therefore, we have reduced the second-order of equation (3-21.1) to the first-order of equation (3-21.7) by the use of the particular solution y = x.

3-3. Homogeneous linear equation with constant coefficients

The constant coefficients a_i in the linear homogeneous differential equation

$$\left(\sum_{i=1}^{n} a_i D^i \right) y(x) = 0 \qquad (3\text{-}22.1)$$

imply the particular solution of the exponential form

$$y(x) = e^{mx} \qquad (3\text{-}22.2)$$

Since the repetitive differential operation on the exponential terms with constant factor in its exponent will yield the constant coefficients a_i

Substituting by y and its derivatives from (3-22.2) in (3-22.1), we get

$$\left(\sum_{i=1}^{n} a_i m^i \right) e^{mx} = 0 \qquad (3\text{-}22.3)$$

3-3.1. The characteristic or auxiliary equation of the linear homogeneous o.d.e.

Since e^{mx} does not vanish, therefore, the bracketed term or the **characteristic** or **auxiliary equation** of the linear homogeneous equation (3-22.1) vanishes as follows

$$\sum_{i=1}^{n} a_i m^i = 0 \qquad (3\text{-}22.4)$$

The roots of the characteristic equation (3-22.4), which determine the values of the n-solutions of the homogeneous linear differential equation, could be one among the following cases:

1. Real and distinct.
2. Real and equal
3. Complex and distinct.
4. Complex and equal.

3-3.2. Real and distinct roots of the characteristic or auxiliary equation

If all roots m_1, m_2, ..m_n, of the characteristic equation (3-22.4) are real and distinct, then all n roots are linearly independent, such that

$$y(x) = \sum_{i=1}^{n} C_i e^{m_i x} \neq 0 \qquad (3\text{-}23)$$

Example 39

Find the characteristic equations and solution of the differential equation

$$\frac{d^2 y}{dx^2} - 3\frac{dy}{dx} + 2y = 0 \qquad (3\text{-}24.1)$$

Solution

Write equation (3-24.1) in terms of the D operator

$$\left(D^2 - 3D + 2\right)y = 0 \qquad (3\text{-}24.2)$$

The auxiliary or characteristic equation is

$$m^2 - 3m + 2 = 0 \qquad (3\text{-}24.3)$$

It roots are

$$m_1 = 1, \qquad m_2 = 2 \qquad (3\text{-}24.4)$$

Thus, from (3-23), the general solution is

$$y(x) = C_1 e^x + C_2 e^{2x} \qquad (3\text{-}24.5)$$

Example 40

Find the characteristic equations and solution of the differential equation

$$\frac{d^3 y}{dx^3} - \frac{dy}{dx} = 0 \qquad (3\text{-}25.1)$$

Equation (3-25.1) is the same equation (3-15.1a) which was solved by the substitution (3-15.2) and integrated over three steps. The following solution is based on the characteristic equation (3-22.4).

Solution

Write equation (3-25.1) in terms of the D operator

$$\left(D^3 - D\right)y = 0 \tag{3-25.2}$$

The auxiliary or characteristic equation is

$$(m+1)(m-1)m = 0 \tag{3-25.3}$$

It roots are

$$m_1 = 0, \qquad m_2 = 1, \qquad m_3 = -1 \tag{3-25.4}$$

Thus, from (3-23), the general solution is

$$y(x) = C_1 + C_2 e^x + C_3 e^{-x} \tag{3-25.5}$$

3-3.3. Real and equal roots of the characteristic or auxiliary equation

If two roots of the characteristic equation (3-22.4) are equal as
$$m_1 = m_2 = k \tag{3-26.1}$$

Then the two solutions

$$e^{m_1 x}, \ e^{kx} \tag{3-26.2}$$

Are linearly dependent particular solutions and do not comprise the general solution.

We will assume that a linearly independent solution is related to the two obtained solutions in (3-26.2) in the form

$$y_2(x) = f(x)e^{kx} \tag{3-26.3}$$

Where, $f(x)$ is unknown function that satisfies the linear homogeneous o.d.e.

Example 41

Find the characteristic equations and solution of the differential equation

$$\frac{d^2y}{dx^2} - 2k\frac{dy}{dx} + k^2 y = 0 \tag{3-27.1}$$

Solution

Write equation (3-27.1) in terms of the D operator

$$\left(D^2 - 2kD + k^2\right)y = 0 \qquad (3\text{-}27.2)$$

The auxiliary or characteristic equation is

$$m^2 - 2km + k^2 = 0 \qquad (3\text{-}27.3)$$

It roots are

$$m_1 = m_2 = k \qquad (3\text{-}27.4)$$

We will assume that a linearly independent solution is related to the two obtained solutions in the form

$$y_2(x) = f(x)e^{kx} \qquad (3\text{-}27.5)$$

Substituting by y_2 and its derivatives from (3-27.5) in equation (3-27.1), we get

$$\left[k^2 e^{kx}f(x) + e^{kx}f''(x) + 2ke^{kx}f'(x)\right] - 2k\left[ke^{kx}f(x) + e^{kx}f'(x)\right] + k^2 e^{kx}f(x) = 0 \qquad (3\text{-}27.6)$$

Arranging, we get

$$e^{kx}f''(x) = 0 \qquad (3\text{-}27.7)$$

Integrating twice, we get

$$\begin{aligned} f'(x) &= C \\ f(x) &= xC \end{aligned} \qquad (3\text{-}27.8)$$

Therefore, the general solution comprises of linear combination of the solution given by root m = k, (3-27.4) and its multiplied form (3-27.5) as follows

$$y(x) = C_1 e^{kx} + C_2 xe^{kx} \qquad (3\text{-}27.9)$$

Example 42

Find the characteristic equations and solution of the differential equation

$$\frac{d^3y}{dx^3} - 3\frac{d^2y}{dx^2} + 3\frac{dy}{dx} - y = 0 \qquad (3\text{-}28.1)$$

Solution

99

Write equation (3-28.1) in terms of the D operator

$$\left(D^3 - 3D^2 + 3D - 1\right)y = 0 \tag{3-28.2}$$

The auxiliary or characteristic equation is

$$m^3 - 3m^2 + 3m - 1 = 0 \tag{3-28.3a}$$

Or,

$$(m-1)^3 = 0 \tag{3-28.3b}$$

It triple root is

$$m_1 = m_2 = m_3 = 1 \tag{3-28.4}$$

We will assume that a linearly independent solution is related to the two obtained solutions in the form

$$y_2(x) = f(x)e^x \tag{3-28.5}$$

Substituting by y_2 and its derivatives from (3-28.5) in equation (3-28.1), we get

$$\left[e^x f(x) + 3e^x f''(x) + 3e^x f'(x) + e^x f'''(x)\right] \\ - 3\left[e^x f(x) + e^x f''(x) + 2e^x f'(x)\right] + 3\left[e^x f(x) + e^x f'(x)\right] - e^x f(x) = 0 \tag{3-28.6}$$

Arranging, we get

$$f'''(x) = 0 \tag{3-28.7}$$

Integrating thrice, we get

$$\begin{aligned} f''(x) &= C \\ f'(x) &= xC + A \\ f(x) &= \frac{x^2}{2}C + Ax + B \end{aligned} \tag{3-28.8}$$

Therefore, the general solution comprises of linear combination of the solution given by root $m = 1$, (3-2.4) and its multiplied form (3-28.5) as follows

$$y(x) = C_1 e^{kx} + \left(\frac{x^2}{2}C + Ax + B\right)e^{kx} \tag{3-28.9}$$

Therefore, the general solution takes the form

$$y(x) = \left(a + bx + cx^2\right)e^{kx} \qquad (3-28.10)$$

3-3.4. Complex roots of the characteristic or auxiliary equation

Since the coefficients a_i in the auxiliary equation (3-22.4), therefore, the characterize roots m's must appear in **conjugate pairs** such that their solution pair adds up to real quantity by canceling the imaginary parts as follows.

$$m_1 = \alpha + i\beta, \qquad m_2 = \alpha - i\beta, \qquad (3-29.1)$$

And conjugate-pair solution

$$y(x) = C_1 e^{(\alpha + i\beta)x} + C_2 e^{(\alpha - i\beta)x} \qquad (3-29.2)$$

The complex exponential terms are expressed by **Euler's formula** (the relationship between the trigonometric functions and the complex exponential function) as follows

$$\begin{aligned} y(x) &= C_1 e^{\alpha x}\left(\cos\beta x + i\sin\beta x\right) + C_2 e^{\alpha x}\left(\cos\beta x - i\sin\beta x\right) \\ &= e^{\alpha x}\cos\beta x\left(C_1 + C_2\right) + i e^{\alpha x}\sin\beta x\left(C_1 - C_2\right) \end{aligned} \qquad (3-29.3)$$

Thus, setting $C_1 = C_2$, we get

$$y_1(x) = Ce^{\alpha x}\cos\beta x \qquad (3-29.4)$$

Since, α and β are arbitrary, therefore, equation (3-29.2) takes the alternate form

$$y(x) = C_1 e^{(\beta + i\alpha)x} + C_2 e^{(\beta - i\alpha)x} \qquad (3-29.5)$$

Hence, we get a second solution in the form

$$y_2(x) = Ae^{\alpha x}\sin\beta x \qquad (3-29.6)$$

Therefore, the linear combination of y_1 and y_2 comprises the general solution

$$y(x) = Ce^{\alpha x}\cos\beta x + Ae^{\alpha x}\sin\beta x \qquad (3-29.7)$$

The two linearly combined solutions are apparently independent since no nonvanishing values of A and B could render $y(x)$ vanish other than in a single point in the range $a \le x \le b$.

Example 43

Find the characteristic equations and solution of the differential equation

$$\frac{d^2y}{dx^2} + 4\frac{dy}{dx} + 5y = 0 \qquad (3\text{-}30.1)$$

Solution

Write equation (3-30.1) in terms of the D operator

$$\left(D^2 + 4D + 5\right)y = 0 \qquad (3\text{-}30.2)$$

The auxiliary or characteristic equation is

$$m^2 + 4m + 5 = 0 \qquad (3\text{-}30.3)$$

It roots are

$$m_1 = -2 + i, \qquad m_2 = -2 - i \qquad (3\text{-}30.4)$$

Thus, from (3-29.7), the general solution is

$$y(x) = e^{-2x}\left(A\cos x + B\sin x\right) \qquad (3\text{-}30.5)$$

Where, A and B are arbitrary constants.

Example 44

Find the characteristic equations and solution of the differential equation

$$\left(D^2 + a^2\right)y = 0 \qquad (3\text{-}31.1)$$

Solution

The auxiliary or characteristic equation is

$$m^2 + a^2 = 0 \qquad\qquad (3\text{-}31.2)$$

It roots are

$$m_1 = ia, \qquad m_2 = -ia \qquad\qquad (3\text{-}31.3)$$

Thus, from (3-29.7), the general solution is

$$y(x) = e^{0x}\left(A\cos ax + B\sin ax\right)$$
$$= A\cos ax + B\sin ax \qquad\qquad (3\text{-}31.4)$$

The two linearly combined solutions are apparently independent since no nonvanishing values of A and B could render y(x) vanish other than in a single point in the interval $a \le x \le b$.

3-3.5. Multiple complex roots of the characteristic or auxiliary equation

Equal complex roots can be treated like equal real roots yet by applying the hypothesis in equation (3-27.5) on both real and imaginary parts of (3-29.5) as many as the **multiplicity** r of the complex roots.

Thus, for a number r of multiple complex roots of the characteristic equation (3-22.4), equation (3-29.7) becomes

$$y(x) = e^{\alpha x}\left\{ \begin{array}{l} \cos\beta x\left[A_0 + A_1 x + A_2 x^2 + .. + A_{r-1}x^{r-1}\right] \\ + \sin\beta x\left[B_0 + B_1 x + B_2 x^2 + .. + B_{r-1}x^{r-1}\right] \end{array} \right\} \qquad (3\text{-}32)$$

Where r is the **multiplicity** of roots

Example 45

Find the characteristic equations and solution of the differential equation

$$\left(D^4 + 2D^2 + 1\right)y = 0 \qquad\qquad (3\text{-}33.1)$$

Solution

The auxiliary or characteristic equation is

$$m^4 + 2m^2 + 1 = 0 \qquad (3\text{-}33.2a)$$

Or,

$$\left(m^2 + 1\right)^2 = 0 \qquad (3\text{-}33.2b)$$

It double roots are

$$m = \pm i \qquad (3\text{-}33.3)$$

Thus, from (3-32), the general solution is

$$y(x) = e^{0x}\{\cos x[A_0 + A_1 x] + \sin x[B_0 + B_1 x]\} \qquad (3\text{-}33.4)$$

3-4. Non-homogeneous linear equation

The non-homogeneous equation (3-1) with **continuous coefficients** a_n in the interval $a \leq x \leq b$ has a unique solution that satisfies the conditions

$$\frac{d^n y(x_0)}{dx^n} = \frac{d^n y_0}{dx^n} \qquad (3\text{-}34.1)$$

Where, x_0 is any point in the interval $a \leq x \leq b$ and $\dfrac{d^n y_0}{dx^n}$ are real numbers.

The solution of non-homogeneous equation has the same linear properties of homogeneous solutions as follows:

1. Factorization

$$\left(\sum a_n(x)D^n\right)Cy(x) = C\left(\sum a_n(x)D^n\right)y(x) = f(x) \qquad (3\text{-}34.2)$$

2. Sum of solutions of homogeneous and non-homogeneous equations

If y_1 is a solution of the **homogeneous** equation

$$\left(\sum a_n(x)D^n\right)y_1(x) = 0 \tag{3-34.3}$$

And if y_2 is a solution of **non-homogeneous** equation (3-1)

$$\left(\sum a_n(x)D^n\right)y_2(x) = f(x) \tag{3-34.4}$$

Adding the two equations (3-34.3) and (3-34.4) by members and noting the properties of the linear operator D, it follows that

$$\left(\sum a_n(x)D^n\right)y_1(x) + \left(\sum a_n(x)D^n\right)y_2(x) = 0 + f(x)$$

$$\left(\sum a_n(x)D^n\right)[y_1(x) + y_2(x)] = f(x) \tag{3-34.5}$$

Thus, y_1+y_2 comprises a solution of the non-homogeneous equation (3-1)

3. Sum of solutions of non-homogeneous equations

If y_i is a solution of the **non-homogeneous** equation (3-34.4). Therefore, a linear combination of the solution y_i comprises a solutions of non-homogeneous equation **provided that its RHS terms is summed in the same factorized manner as the solution y_i,** as follows

$$\left(\sum a_n(x)D^n\right)y_i(x) = f_i(x) \tag{3-34.6}$$

Multiplying both sides by the same arbitrary constant Ci and adding, noting the linear property of the differential operator D, we get

$$\sum_{i=1}^{n}\left(\sum a_n(x)D^n\right)C_i y_i(x) = \sum_{i=1}^{n} C_i f_i(x)$$

$$\left(\sum a_n(x)D^n\right)\sum_{i=1}^{n} C_i y_i(x) = \sum_{i=1}^{n} C_i f_i(x) \tag{3-34.7}$$

Note that the linear combination of the solutions off the non-homogeneous equation **is not a solution of the initial non-homogeneous equation** but a solution of an altered equation with the RHS of the non-homogenous equation summed in the same manner of linear combination of the solutions.

4. Particular and Complementary functions of non-homogeneous

105

From the properties of homogeneous equation, equation (3-34.5) and those of non-homogeneous equation (3-34.7), we conclude the following:

(a) There is only one **particular solution** that satisfies the non-homogeneous equation (3-34.6), which is **devoid of arbitrary constants**.

(b) There is a sum of linear combinations of solutions of the homogenous equation (3-34.3), which contains a numbers of **arbitrary constants** equal to the number of those solutions and which comprises the **complementary function** of the homogenous part of the equation. .

(c) The **general solution** of non-homogeneous equation comprises of the sum of **complementary function** and the **particular solution**, as follows:

$$y(x) = y_{P.S.} + \sum C_i y_i$$

(3-34.8)

$$G.S. = P.S. + \quad C.S.$$

Example 46

Find the general and particular solutions of the non-homogenous differential equation

$$(D^2 + 1)y = x$$

(3-35.1)

Solution

(a) **Particular solution** of the non-homogenous equation (3-35.1) is

$$y_1 = x$$

(3-35.2)

Satisfies the non-homogeneous equation (3-35.1).

(b) **Complementary function** of the homogenous equation

$$(D^2 + 1)y = 0$$

(3-35.3)

The auxiliary or characteristic equation is

$$m^2 + 1 = 0$$

(3-35.4)

It double roots are

$$m = \pm i$$

(3-35.5)

Thus, from (3-32), the general solution of equation (3-35.3) is

$$y(x) = A\cos x + B\sin x \qquad (3\text{-}35.6)$$

(c) **General Solution** = Complementary function + Particular Solution

Equations (3-35.2) and (3-35.6) yield the general solution

$$y(x) = x + A\cos x + B\sin x \qquad (3\text{-}35.7)$$

3-5. Non-homogeneous linear equation with constant coefficients

The constant coefficients of a non-homogeneous linear equation render the identification of particular solution an easy task as follows:

(a) RHS is **polynomial** of nth degree
(b) RHS is polynomial of nth degree multiplied by **exponential** function of x
(c) RHS is polynomial of nth degree multiplied by **complex** exponential function of x

3-5.1. Non-homogeneous linear o.d.e. with polynomial function on the RHS

Given the linear homogenous o.d.e. with constant coefficients a_i, which RHS is a polynomial of s-degree, as follows

$$\left(\sum_{i=0}^{n} a_i D^i\right) y(x) = \sum_{j=0}^{j=s} b_j x^j \qquad (3\text{-}36.1)$$

A **particular solution** of the form

$$y(x) = \sum_{j=0}^{j=s} c_j x^j \qquad (3\text{-}36.2)$$

Where the constants c_j are determined by substituting, by y and its derivatives, from (3-36.2), in the LHS of equation (3-36.1), then equating the coefficients of **equal powers** of x on both sides to determines c_j in terms of a_i and b_j.

Therefore, if $a_0 \neq 0$, there exists a particular solution (integral) of the form (3-36.2) of the same degree of the polynomial on the RHS of equation (3-36.1).

Example 47

Find the particular and general solutions for the o.d.e.

$$y''(x) + y(x) = x^2 + x \tag{3-37.1}$$

Solution

(a) The **particular solution** has the degree of the polynomial on the RHS of (3-37.1) as follows

$$y(x) = c_2 x^2 + c_1 x + c_0 \tag{3-37.2}$$

In order to determine the constants c's, substitute by y and its derivatives from (3-37.2) in (3-37.1) to get

$$(2c_2) + (c_2 x^2 + c_1 x + c_0) = x^2 + x \tag{3-37.3}$$

Equating the coefficients of equal powers of x, we get

$$\begin{aligned} c_2 &= 1 \\ c_1 &= 1 \\ c_0 &= -2 \end{aligned} \tag{3-37.4}$$

Thus, the P.S. is

$$y(x) = x^2 + x - 2 \tag{3-37.5}$$

(b) The **complementary function** of the homogeneous equation is obtained as before as follows:

$$y''(x) + y(x) = 0 \tag{3-37.6}$$

The characteristic equation is

$$m^2 + 1 = 0 \tag{3-37.7}$$

Thus, the C.F. is

$$y(x) = A\cos x + B\sin x \tag{3-37.8}$$

(c) The **general solution** of the non-homogeneous equation is the sum of P.S., (3-37.5), and C.F., (3-37.8), as follows:

$$y(x) = C.F. + P.S.$$
$$= A\cos x + B\sin x + x^2 + x - 2$$

(3-37.9)

Example 48

Find the particular and general solutions for the o.d.e.

$$y''(x) + y'(x) = x - 2 \qquad (3\text{-}38.1)$$

Solution

(a) Let us first assume that the **particular solution** has the polynomial on the RHS of (3-38.1) as follows

$$y(x) = c_2 x^2 + c_1 x + c_0 \qquad (3\text{-}38.2)$$

In order to determine the constants c's, substitute by y and its derivatives from (3-38.2) in (3-38.1) to get

$$(2c_2) + (2c_2 x + c_1) = x - 2 \qquad (3\text{-}38.3)$$

Equating the coefficients of equal powers of x, we get

$$c_2 = \frac{1}{2}$$
$$c_1 = -3 \qquad (3\text{-}38.4)$$
$$c_0 = ?$$

We note that c_0 remains undetermined since equation (3-38.1) is devoid from the constant a_0, the coefficient of y.

Therefore, in case that a0 is absent on the LHS, it suffices to propose a particular solution of the form

$$y(x) = x(c_2 x + c_1) \qquad (3\text{-}38.5)$$

Thus, the P.S. is

$$y(x) = x\left(\frac{1}{2}x - 3\right) \qquad (3\text{-}38.6)$$

(b) The **complementary function** of the homogeneous equation is obtained as before as follows:

$$y''(x) + y'(x) = 0 \tag{3-38.7}$$

The characteristic equation is

$$m^2 + m = 0 \tag{3-38.8}$$

Thus, the C.F. is

$$y(x) = A + Be^{-x} \tag{3-38.9}$$

(c) The **general solution** of the non-homogeneous equation is the sum of P.S., (3-38.6), and C.F., (3-38.9), as follows:

$$y(x) = C.F. + P.S.$$
$$= A + Be^{-x} + x\left(\frac{1}{2}x - 3\right) \tag{3-38.10}$$

3-5.2. Non-homogeneous linear o.d.e. with polynomial function in x multiplied by exponential function, on the RHS

$$\left(\sum_{i=0}^{n} a_i D^i\right) y(x) = e^{\beta x} \sum_{j=0}^{j=s} b_j x^j \tag{3-39.1}$$

A **particular solution** of the form

$$y(x) = e^{\beta x} \sum_{j=0}^{j=s} c_j x^j \tag{3-39.2}$$

Satisfies equation (3-39.1) when a_0 does not vanish, similar to equation (3-36.2). Clearly, the exponential function in (3-39.2) transforms equation (3-39.1)

A **particular solution** of the form

$$y(x) = e^{\beta x} \sum_{j=0}^{j=s} c_j x^j \tag{3-39.3}$$

Satisfies equation (3-39.1) when a_0 does not vanish, similar to equation (3-36.2). Clearly, the exponential function in (3-39.3) transforms equation (3-39.2) onto a linear homogeneous equation with constant coefficients.

110

Also, similar to equation (3-38.5), when a_0 vanishes, the particular solution takes the form

$$y(x) = x^r e^{\beta x} \sum_{j=0}^{j=s} c_j x^j \qquad (3-39.4)$$

Example 49

Find the particular and general solutions for the o.d.e.

$$\left(D^2 + 9\right)y = e^{5x} \qquad (3-40.1)$$

Solution

(a) Let us first assume that the **particular solution** has the form of the RHS of (3-40.1) as follows

$$y = Ce^{5x} \qquad (3-40.2)$$

In order to determine the constants c's, substitute by y and its derivatives from (3-40.2) in (3-40.1) to get

$$\left(25C\right) + 9C = 1 \qquad (3-40.3a)$$

Thus,

$$C = 1/34 \qquad (3-40.3b)$$

(b) The **complementary function** of the homogeneous equation is obtained as before as follows:

$$y''(x) + 9y(x) = 0 \qquad (3-40.4)$$

The characteristic equation is

$$m^2 + 9 = 0 \qquad (3-40.5)$$

Thus, the C.F. is

$$y(x) = A\cos 3x + B\sin 3x \qquad (3-40.6)$$

(c) The **general solution** of the non-homogeneous equation is the sum of P.S., (3-40.2), and C.F., (3-40.6), as follows:

111

$$y(x) = C.F. + P.S.$$

$$= A\cos 3x + B\sin 3x + \frac{1}{34}e^{5x} + \qquad (3\text{-}40.7)$$

Example 50

Find the particular and general solutions for the o.d.e.

$$\left(D^2 + 1\right)y = e^{3x}(x - 2) \qquad (3\text{-}41.1)$$

Solution

(a) Let us first assume that the **particular solution** has the form of the RHS of (3-41.1) as follows

$$y = e^{3x}(c_1 x + c_0) \qquad (3\text{-}41.2)$$

In order to determine the constants c's, substitute by y and its derivatives from (3-41.2) in (3-41.1) to get

$$16e^{3x}(c_1 x) + (16e^{3x}c_0) = e^{3x}(x - 2) \qquad (3\text{-}41.3a)$$

$$c_1 = \frac{1}{16}$$
$$c_0 = -\frac{1}{8} \qquad (3\text{-}41.3b)$$

i.e., $\qquad y = \frac{1}{16}e^{3x}(x - 2) \qquad (3\text{-}41.4)$

(b) The **complementary function** of the homogeneous equation is obtained as before as follows:

$$y''(x) + y(x) = 0 \qquad (3\text{-}41.5)$$

The characteristic equation is

$$m^2 + 1 = 0 \qquad (3\text{-}41.6)$$

Thus, the C.F. is

$$y(x) = A\cos x + B\sin x \qquad (3\text{-}41.7)$$

(c) The **general solution** of the non-homogeneous equation is the sum of P.S., (3-40.2), and C.F., (3-40.6), as follows:

112

$$y(x) = \text{C.F.} + \text{P.S.}$$
$$= A\cos x + B\sin x + \frac{1}{16}e^{3x}(x-2) + \qquad (3\text{-}41.8)$$

Example 51

Find the particular and general solutions for the o.d.e.

$$\left(D^2 - 1\right)y = e^x(x^2 - 1) \qquad (3\text{-}42.1)$$

Solution

(a) Note that the left side has the exponential function and is devoid of x. Therefore, we assume the following form of **particular solution**

$$y = xe^x(c_2 x^2 + c_1 x + c_0) \qquad (3\text{-}42.2)$$

The proposition of degree of the polynomial will be apparent was we substitute by it in the differential equation (3-42.1)

In order to determine the constants c's, substitute by y and its derivatives from (3-42.2) in (3-42.1) to get

$$6c_2 x^2 + (4c_1 + 6c_2)x + (2c_0 + 2c_1) = (x^2 - 1) \qquad (3\text{-}42.3a)$$

$$c_2 = \frac{1}{6}$$

$$c_0 = -\frac{1}{4} \qquad (3\text{-}42.3b)$$

$$c_1 = -\frac{1}{4}$$

i.e.,
$$y = xe^x\left(\frac{1}{6}x^2 - \frac{1}{4}x - \frac{1}{4}\right) \qquad (3\text{-}42.4)$$

(b) The **complementary function** of the homogeneous equation is obtained as before as follows:

$$y''(x) - y(x) = 0 \qquad (3\text{-}42.5)$$

The characteristic equation is

$$m^2 - 1 = 0 \qquad (3\text{-}42.6)$$

113

Thus, the C.F. is

$$y(x) = Ae^x + Be^{-x} \tag{3-42.7}$$

(c) The **general solution** of the non-homogeneous equation is the sum of P.S., (3-42.4), and C.F., (3-42.7), as follows:

$$y(x) = \text{C.F.} + \text{P.S.}$$

$$= Ae^x + Be^{-x} + xe^x\left(\frac{1}{6}x^2 - \frac{1}{4}x - \frac{1}{4}\right) \tag{3-42.8}$$

Example 52

Find the particular and general solutions for the o.d.e.

$$\left(D^3 + 3D^2 + 3D + 1\right)y = e^{-x}(x - 5) \tag{3-43.1}$$

Solution

(a) Note that the left side has the exponential function and is devoid of x^3. Therefore, we assume the following form of **particular solution**

$$y = x^3 e^{-x}(c_1 x + c_0) \tag{3-43.2}$$

In order to determine the constants c's, substitute by y and its derivatives from (3-43.2) in (3-43.1) to get

$$
\begin{aligned}
&x^3 e^{-x}(c_1 x + c_0) \\
&+ 3\left(-e^{-x}(c_1 x^4 + x^3 c_0) + e^{-x}(4c_1 x^3 + 3x^2 c_0)\right) \\
&+ 3e^{-x}\left(c_1 x^4 + x^3(c_0 - 8c_1) + (-6c_0 + 12c_1)x^2 + 9xc_0\right) \\
&- e^{-x}\left(c_1 x^4 + x^3(c_0 - 8c_1) + (-6c_0 + 12c_1)x^2 + 9xc_0\right) \\
&+ e^{-x}\left(4c_1 x^3 + 3x^2(c_0 - 8c_1) + 2(-6c_0 + 12c_1)x + 9c_0\right) = e^{-x}(x - 5)
\end{aligned} \tag{3-43.3a}
$$

Arranging, we get

$$24c_1 x + 6c_0 = (x - 5) \tag{3-43.3b}$$

Therefore,

114

$$c_1 = \frac{1}{24}$$

$$c_0 = \frac{-5}{6}$$

(3-43.3c)

Therefore, the particular solution, (3-43.2) becomes

i.e.,
$$y = x^3 e^{-x} \left(\frac{1}{24} x - \frac{5}{6} c_0 \right)$$

(3-43.4)

(b) The **complementary function** of the homogeneous equation is obtained as before as follows:

The characteristic equation is

$$(m+1)^3 = 0$$

(3-43.5)

Thus, the C.F. is obtained in the form of (3-28.5)

$$y(x) = f(x)e^{-x}$$

(3-43.6)

Where, as before, f(x) is determined by substituting by y and its derivatives in (3-43.1) as follows

$$\begin{pmatrix} fe^{-x} \\ + 3(f'e^{-x} - fe^{-x}) \\ + 3(f''e^{-x} - 2f'e^{-x} + fe^{-x}) \\ + (f'''e^{-x} - 3f''e^{-x} + 3f'e^{-x} - fe^{-x}) \end{pmatrix} y = 0$$

(3-43.7)

Arranging we get f''', which is integrated thrice as follows:

$$f''' = 0$$
$$f'' = C$$
$$f' = xC + A$$
$$f = \frac{x^2}{2} C + Ax + B$$

(3-43.8)

Thus, the complementary function, equation (3-43.6), is

$$y(x) = \left(\frac{x^2}{2} C + Ax + B \right) e^{-x}$$

(3-43.9)

(c) The **general solution** of the non-homogeneous equation is the sum of P.S., (3-43.4), and C.F., (3-43.7), as follows:

$$y(x) = C.F. + P.S.$$

$$= \left(\frac{x^2}{2}C + Ax + B \right)e^{-x} + x^3 e^{-x} \left(\frac{1}{24}x - \frac{5}{6}c_0 \right) \tag{3-43.10}$$

3-5.3. Non-homogeneous linear o.d.e. with complex function the RHS

If the RHS is complex, the same above rules of finding the particular solution apply as follows:

$$\left(\sum_{i=0}^{n} a_i D^i \right) y(x) = e^{\alpha + i\beta} \sum_{j=0}^{j=s} b_j x^j + e^{\alpha - i\beta} \sum_{j=0}^{j=s} c_j x^j \tag{3-44.1}$$

(a) If $(\alpha \pm i\beta)$ are not roots of the characteristic equation of the homogeneous part of the above equation (LHS=0), then the particular solution takes the form

$$y(x) = e^{\alpha} \left(\cos\beta x \sum_{j=0}^{j=s} B_j x^j + \sin\beta x \sum_{j=0}^{j=s} C_j x^j \right) \tag{3-44.2}$$

Where the unknown constants B's and C's are determined by substitution in the LHS of (3-44.1).

(b If $(\alpha \pm i\beta)$ are roots of **multiplicity r** of the characteristic equation of the homogeneous part of the above equation (LHS=0), then the particular solution takes the form

$$y(x) = x^r e^{\alpha} \left(\cos\beta x \sum_{j=0}^{j=s} B_j x^j + \sin\beta x \sum_{j=0}^{j=s} C_j x^j \right) \tag{3-44.3}$$

Similarly, the unknown constants B's and C's are determined by substitution in the LHS of (3-44.1).

Example 53

Find the particular and general solutions for the o.d.e.

$$\left(D^2 + 4D + 4 \right)y = \cos 2x \tag{3-45.1}$$

Solution

(a) The **complementary function** of the homogeneous equation is obtained as before as follows:

The characteristic equation is

$$(m+2)^2 = 0 \tag{3-45.2}$$

Its roots are $m_1 = m_2 = -2$.

Thus, the C.F. is obtained in the form of (3-28.5)

$$y(x) = f(x)e^{-x} \tag{3-45.3}$$

Substituting by y and its derivatives from (3-45.3) in the homogeneous part of equation (3-45.1), (LHS=0), we get

$$4fe^{-2x}$$
$$+4\left(-2fe^{-2x} + f'e^{-2x}\right)$$
$$+\left(4fe^{-2x} - 2f'e^{-2x} - 2f'e^{-2x} + f''e^{-2x}\right) = 0$$

Arranging, then integrating,

$$f'' = 0$$
$$f' = C \tag{3-45.4}$$
$$f = Cx + A$$

Thus, the C.F. is

$$y(x) = e^{-x}(Cx + A) \tag{3-45.5}$$

Thus, according to equation (3-34.8), we seek a particular solution that includes the cosine term of ($\pm 2i$) roots, which according to equation (3-39.4), yield the real cosine term.

(b) Therefore, we assume the following form of **particular solution**

$$y = c_1 \cos 2x + c_2 \sin 2x \tag{3-45.6}$$

Where the constants c's are determined by substituting from equation (3-45.6) in (3-45.1), to get

$$4\left(c_1 \cos 2x + c_2 \sin 2x\right)$$
$$+4\left(-2c_1 \sin 2x + 2c_2 \cos 2x\right)$$
$$+\left(-4c_1 \cos 2x - 4c_2 \sin 2x\right) = \cos 2x$$

Or, $$-8c_1 \sin 2x + 8c_2 \cos 2x = \cos 2x \tag{3-45.7}$$

Equating the terms of similar terms from both sides, we get

Or,
$$c_1 = 0$$
$$c_2 = \frac{1}{8}$$
(3-45.8)

Thus, the particular solution becomes

$$y = \frac{1}{8}\sin 2x$$
(3-45.9)

(c) The general solution is obtained from (3-45.5) and (3-45.9) as

G.S. = C.F. + P.S.

$$y(x) = e^{-x}(Cx + A) + \frac{1}{8}\sin 2x$$
(3-45.10)

Example 54

Find the particular and general solutions for the o.d.e.

$$\left(D^4 + 2D^2 + 1\right)y = \sin x$$
(3-46.1)

Solution

(a) Complementary function

The **complementary function** of the homogeneous equation is obtained as before as follows:

The characteristic equation is

$$\left(m^2 + 1\right)^2 = 0$$
(3-46.2)

Its roots are $m_1 = -i$ and $m_2 = +i$, of double **multiplicity**.

Thus, the C.F. is obtained in the form of (3-30.5)

$$y(x) = f(x)\left(Ae^{-ix} + Be^{+ix}\right)$$
(3-46.3)

Substituting by y and its derivatives from (3-46.3) in the homogeneous part of equation (3-46.1), (LHS=0), we get

118

$$f\left(Ae^{-ix} + Be^{+ix}\right)$$
$$+ 2\left(f''\left(Ae^{-ix} + Be^{+ix}\right) + 2f'\left(-iAe^{-ix} + iBe^{+ix}\right) - f\left(Ae^{-ix} + Be^{+ix}\right)\right)$$
$$+ f''''\left(Ae^{-ix} + Be^{+ix}\right) + 4f'''\left(-iAe^{-ix} + iBe^{+ix}\right) - 6f''\left(Ae^{-ix} + Be^{+ix}\right)$$
$$- 4f'\left(-iAe^{-ix} + iBe^{+ix}\right) + f\left(Ae^{-ix} + Be^{+ix}\right) = 0 \tag{3-46.4}$$

Arranging, then integrating,

$$f''''\left(Ae^{-ix} + Be^{+ix}\right) + 4f'''\left(-iAe^{-ix} + iBe^{+ix}\right) - 4f''\left(Ae^{-ix} + Be^{+ix}\right) = 0 \tag{3-46.5}$$

Separating the real and imaginary parts, we get

$$f''''(A + B)\cos x + 4f'''(-A - B)\sin x - 4f''(A + B)\cos x = 0$$
$$f''''(-A + B)\sin x + 4f'''(-A + B)\cos x - 4f''(-A + B)\sin x = 0 \tag{3-46.6}$$

Dividing by the common factors in each equation, we get

$$f''''\cos x - 4f'''\sin x - 4f''\cos x = 0$$
$$f''''\sin x + 4f'''\cos x - 4f''\sin x = 0 \tag{3-46.7}$$

Multiplying first equation by sinx, second by cosx, subtracting, we get

$$f''''\cos x \sin x - 4f'''\sin x \sin x - 4f''\cos x \sin x = 0$$
$$f''''\cos x \sin x + 4f'''\cos x \cos x - 4f''\cos x \sin x = 0 \tag{3-46.8}$$
$$f''' = 0$$

Substituting from (3-46.8) in (3-46.7), we get

$$-4f'''\sin x - 4f''\cos x = 0$$
$$4f'''\cos x - 4f''\sin x = 0 \tag{3-46.9}$$

Multiplying first equation by cosx, second by sinx, adding, we get

$$-4f'''\cos x \sin x - 4f''\cos x \cos x = 0$$
$$4f'''\sin x \cos x - 4f''\sin x \sin x = 0 \tag{3-46.10}$$
$$f'' = 0$$

Integrating twice, we get

$$f'' = 0$$
$$f' = C$$
$$f = Cx + A \tag{3-46.11}$$

Thus, we obtained the form in equation (3-32), where multiplicity r of the imaginary roots is equal 2, rendering the polynomial f=Cx+A, equation (3-46.11).

Thus, the C.F. is

$$y(x) = [A_0 + A_1 x]\cos x + [B_0 + B_1 x]\sin x \qquad (3-46.12)$$

(b) Particular Solution

Thus, according to equation (3-34.8), we seek a particular solution that includes the combinations of cosine and sines, but not included in the C.F., (3-46.12) term of ($\pm 2i$) roots, with the multiplicity $r = 2$ term in equation (3-44.3), as follows:

$$y(x) = x^2 [C_1 \cos x + C_2 \sin x] \qquad (3-46.13)$$

Where the constants C's are determined by substituting from equation (3-46.13) in (3-46.1), to get

$$x^2 [C_1 \cos x + C_2 \sin x]$$
$$+ 2(x^2 [-C_1 \cos x - C_2 \sin x] + 4x[-C_1 \sin x + C_2 \cos x] + 2[C_1 \cos x + C_2 \sin x]) \qquad (3-46.14)$$
$$+ (x^2 [+C_1 \cos x + C_2 \sin x] + 8x[+C_1 \sin x - C_2 \cos x] + 12[-C_1 \cos x - C_2 \sin x]) = \sin x$$

$$8[-C_1 \cos x - C_2 \sin x] = \sin x \qquad (3-46.15)$$

Equating the coefficients of similar functions on both sides we get

$$C_2 = -\frac{1}{8}$$
$$C_1 = 0 \qquad (3-46.16)$$

Thus, the particular solution, (3-46.13), becomes

$$y(x) = -\frac{x^2}{8}\sin x \qquad (3-46.17)$$

(c) The general solution is obtained from (3-46.12) and (3-46.17) as

$$G.S. = C.F. + P.S.$$
$$y(x) = [A_0 + A_1 x]\cos x + [B_0 + B_1 x]\sin x - \frac{x^2}{8}\sin x \qquad (3-46.18)$$

120

Example 55

Find the particular and general solutions for the o.d.e.

$$\left(D^2 + 2D + 2\right)y = e^{-x}(x\cos x + 3\sin x)$$ (3-47.1)

Solution

(a) Complementary function

The **complementary function** of the homogeneous equation is obtained as before as follows:

The characteristic equation is

$$m^2 + 2m + 2 = 0$$ (3-47.2)

Its roots are ($m_1 = -1-i$) and ($m_2 = -1+i$), of single **multiplicity**.

Thus, the C.F. is obtained in the form of (3-39.7)

$$y(x) = e^{-x}\left(A\cos x + B\sin x\right)$$ (3-47.3)

(b) Particular Solution

Since, the complementary function, equation (3-47.3), contains the products of sines and cosines by the exponential function in x, the particular solution of equation (3-47.1) takes the form (3-44.3), with multiplicity $r = 1$.

$$y(x) = x^r e^{\alpha}\left(\cos\beta x \sum_{j=0}^{j=s} B_j x^j + \sin\beta x \sum_{j=0}^{j=s} C_j x^j\right)$$ (3-47.4)

$$= xe^{-x}\left[(C_1 x + C_0)\cos x + (C_3 x + C_2)\sin x\right]$$

Where the constants C's are determined by substituting by y and its derivatives from equation (3-47.4) in (3-47.1), as follows.

The derivatives of y, equation (3-47.4):

$$y' = e^{-x}\left[\begin{array}{l} \cos x\left[x(2C_1 + C_2 - C_0) + (C_3 - C_1)x^2 + C_0\right] \\ + \sin x\left[x(2C_3 - C_0 - C_2) - (C_3 + C_1)x^2 + C_2\right]\end{array}\right] \qquad (3\text{-}47.5a)$$

$$y'' = 2e^{-x}\left[\begin{array}{l} \sin x\left[C_1 x^2 - x(2C_1 + 2C_3 - C_0) - (C_0 + C_2 - C_3)\right] \\ - \cos x\left[C_3 x^2 - x(2C_3 - C_2 - 2C_1) - (C_2 + C_1 - C_0)\right]\end{array}\right] \qquad (3\text{-}47.5b)$$

Substituting from (3-47.5) into the initial equation (3-47.1), we get

$$2x\left[(C_1 x + C_0)\cos x + (C_3 x + C_2)\sin x\right]$$
$$+ 2\left[\begin{array}{l}\cos x\left[x(2C_1 + C_2 - C_0) + (C_3 - C_1)x^2 + C_0\right] \\ + \sin x\left[x(2C_3 - C_0 - C_2) - (C_3 + C_1)x^2 + C_2\right]\end{array}\right]$$
$$+ 2\left[\begin{array}{l}+ \sin x\left[C_1 x^2 - x(2C_1 + 2C_3 - C_0) - (C_0 + C_2 - C_3)\right] \\ - \cos x\left[C_3 x^2 - x(2C_3 - C_2 - 2C_1) - (C_2 + C_1 - C_0)\right]\end{array}\right]$$
$$= (x\cos x + 3\sin x)$$

Arranging, we get

$$\left[\begin{array}{l}\cos x\left[4xC_3 + 2(C_2 + C_1)\right] \\ + \sin x\left[-4xC_1 - 2(C_0 - C_3)\right]\end{array}\right] = (x\cos x + 3\sin x) \qquad (3\text{-}47.5c)$$

On the LHS, we have the four terms cosx, sin x, x cosx, x sinx.
On the RHS, we have the two terms x cosx and sin x.

Equating the coefficients of similar terms on both sides, we get

$$C_3 = \frac{1}{4}$$
$$C_2 = C_1 = 0 \qquad (3\text{-}47.6)$$
$$C_0 = -\frac{5}{4}$$

Putting the C's in equation (3-47.4), we get

$$y(x) = xe^{-x}\left[-\frac{5}{4}\cos x + \frac{x}{4}\sin x\right] \qquad (3\text{-}47.7)$$

(c) General Solution

122

G.S. = C.F. + P.S.

$$y(x) = xe^{-x}\left[-\frac{5}{4}\cos x + \frac{x}{4}\sin x\right] + e^{-x}(A\cos x + B\sin x) \qquad (3\text{-}47.8)$$

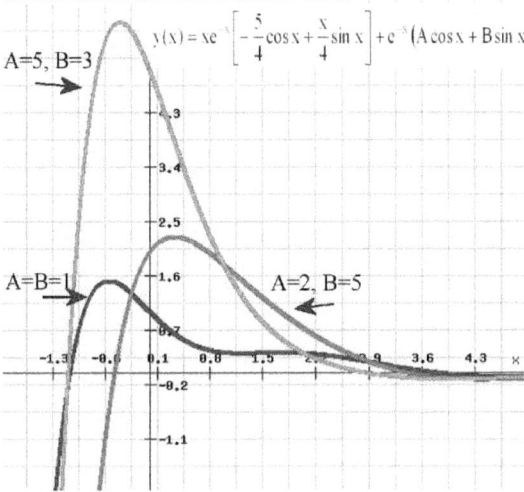

f(x)=(1/e^x)*(cos(x)+sin(x))+(x/4)*(1/e^x)*(x*sin(x)-5*cos(x))
g(x)=(1/e^x)*(2*cos(x)+5*sin(x))+(x/4)*(1/e^x)*(x*sin(x)-5*cos(x))
h(x)=(1/e^x)*(5*cos(x)+5*sin(x))+(x/4)*(1/e^x)*(x*sin(x)-5*cos(x))

$$y(x) = xe^{-x}\left[-\frac{5}{4}\cos x + \frac{x}{4}\sin x\right] + e^{-x}(A\cos x + B\sin x)$$

A=5, B=3

A=B=1

A=2, B=5

Example 56

Find the particular solutions for the o.d.e. using Euler's formula to transform cosx to exponential function.

$$(D^2 - 2D + 1)y = \cos x \qquad (3\text{-}48.1)$$

Solution

Using Euler's formula. we get

$$(D^2 - 2D + 1)y = e^{ix} \qquad (3\text{-}48.2)$$

The particular solution of equation (3-48.2) may take the form

$$y = Ce^{ix} \qquad (3\text{-}48.3)$$

Where the constants C is determined by substituting by y and its derivatives from equation (3-48.3) in (3-48.1), as follows.

$$-Ce^{ix} - 2iCe^{ix} + Ce^{ix} = e^{ix}$$
$$-C - 2iC + C = 1 \tag{3-48.4}$$
$$C = \frac{i}{2}$$

Therefore, the particular solution, (3-48.3) becomes

$$y = \frac{i}{2}e^{ix}$$
$$= \frac{1}{2}(i\cos x - \sin x) \tag{3-48.5}$$

The real part

$$y = -\frac{1}{2}\sin x \tag{3-48.6}$$

3-6. Inverse D-Operator

1. Integrator Operator

$$\int f(x)dx = \frac{1}{F(D)}f(x) \tag{3-49.1}$$

2. Integrator –Differential Operator

$$f(x) = \frac{1}{F(D)}[F(D)f(x)] \tag{3-49.2}$$

3. Distributive Integrator –Differential Operator

$$G(D)f(x) = G(D)\left[\frac{1}{F(D)}f(x)\right]$$

$$= \frac{1}{F(D)}[G(D)f(x)]$$

(3-49.3)

4. Multiple Integrator Operator

$$\frac{1}{D^n}f(x) = \int ..\left(\int\left(\int f(x)dx\right)dx\right)dx$$

$$= \int\int\int...\int\left(\int f(x)dx^n\right)$$

(3-49.4)

5. Factorization of the Inverse D-Operator

From equation (3-4.1), we can prove that

$$\frac{1}{F(D)}Cf(x) = C\frac{1}{F(D)}f(x) = Cy(x)$$

(3-49.5)

6. Integration of exponential function

From equation (3-16.2), we can prove that

$$\frac{1}{F(D)}e^{\alpha x} = \frac{e^{\alpha x}}{F(\alpha)}, \qquad F(\alpha) \neq 0$$

(3-49.6)

7. Double Integration of sine function

From equation (3-17.2), we can prove that

$$\frac{1}{F(D^2)}\sin\beta x = \frac{1}{F(-\beta^2)}\sin\beta x$$

(3-49.7)

8. Double Integration of cosine function

From equation (3-18.2), we can prove that

$$\frac{1}{F(D^2)}\cos\beta x = \frac{1}{F(-\beta^2)}\cos\beta x \qquad\qquad (3\text{-}49.8)$$

9. Integration of a product of a function by exponential function

From equation (3-19.1), we have

$$F(D)e^{\alpha x}g(x) = e^{\alpha x}F(D+\alpha)g(x) \qquad\qquad (3\text{-}49.9a)$$

Replace the arbitrary function $g(x)$ by $h(x)$ as follows

$$h(x) = \frac{g(x)}{F(D+\alpha)} \qquad\qquad (3\text{-}49.9b)$$

Therefore, we have (3-49.9a) becomes

$$F(D)e^{\alpha x}h(x) = e^{\alpha x}F(D+\alpha)h(x)$$
$$= e^{\alpha x}F(D+\alpha)\frac{g(x)}{F(D+\alpha)} \qquad\qquad (3\text{-}49.9c)$$
$$= e^{\alpha x}g(x)$$

Apply the differential-integrator on both sides of the above equation, from equation (3-49.2), we get

$$\frac{1}{F(D)}e^{\alpha x}g(x) = \frac{1}{F(D)}F(D)e^{\alpha x}h(x)$$
$$= e^{\alpha x}h(x) \qquad\qquad (3\text{-}49.9d)$$
$$= e^{\alpha x}\frac{g(x)}{F(D+\alpha)}$$

Therefore, the integration of a function multiplied by an exponential function **shifts the differential operator D by the factor α of the exponent of the exponential function.**

126

10. Distributive Inverse D-Operator

From equation (3-7.3), we can prove that

$$\frac{1}{F(D)}\big((f(x)+f(g)\big)=\frac{1}{F(D)}f(x)+\frac{1}{F(D)}g(x) \qquad (3\text{-}49.10)$$

11. Association of Inverse D-Operator

From equation (3-7.2), we can prove that

$$\frac{1}{F(D)H(D)}f(x)=\frac{1}{F(D)}\left(\frac{1}{H(D)}f(x)\right) \qquad (3\text{-}49.11)$$

Example 57

Integrate the following o.d.e. using the Inverse D-Operator

$$\frac{d^2y}{dx^2}+4y=e^x \qquad (3\text{-}50.1)$$

Solution

The D-Operator form is

$$\left(D^2+4\right)y=e^x \qquad (3\text{-}50.2)$$

Equation (3-49.3) allows us to write

$$y=\frac{1}{\left(D^2+4\right)}e^x \qquad (3\text{-}50.3)$$

(a) Decomposing it into partial fractions, we get

$$y = \frac{1}{D^2 - (2i)^2} e^x$$

$$= \frac{1}{(D + 2i)(D - 2i)} e^x \qquad (3\text{-}50.4)$$

$$= \frac{1}{4i} \left(\frac{1}{(D - 2i)} - \frac{1}{(D + 2i)} \right) e^x$$

Equation (3-49.9d) allows us to write

$$y = \frac{1}{4i} \left(e^{2ix} \frac{1}{D} e^{(1-2i)x} - e^{-2ix} \frac{1}{D} e^{(1+2i)x} \right) \qquad (3\text{-}50.5)$$

Integrating, we get

$$y = \frac{1}{4i} \left(e^{2ix} \frac{1}{(1 - 2i)} e^{(1-2i)x} - e^{-2ix} \frac{1}{(1 + 2i)} e^{(1+2i)x} \right)$$

$$= \frac{1}{4i} \left(\frac{4i}{(1 - 2i)(1 + 2i)} \right) e^x \qquad (3\text{-}50.6)$$

$$= \frac{1}{5} e^x$$

Which is the particular solution (3-50.1)

(b) Alternative solution

Another solution comes through equation (3-49.6) as follows

$$y = \frac{1}{(1^2 + 4)} e^x = \frac{1}{5} e^x \qquad (3\text{-}50.7)$$

Example 58

Integrate the following o.d.e. using the Inverse D-Operator

$$\frac{d^4 y}{dx^4} + y = 2\cos 3x \qquad (3\text{-}51.1)$$

Solution

The D-Operator form is

$$\left(D^4 + 1\right)y = 2\cos 3x \qquad (3\text{-}51.2)$$

Equation (3-49.3) allows us to write

$$y = \frac{2}{\left(D^4 + 1\right)}\cos 3x \qquad (3\text{-}51.3)$$

(a) Decomposing it into partial fractions, we get

$$y = \frac{2}{D^4 - (i)^2}\cos 3x$$
$$= \frac{2}{\left(D^2 + i\right)\left(D^2 - i\right)}\cos 3x \qquad (3\text{-}51.4)$$

Equation (3-49.8) gives the double integration of the cosine

$$y = \frac{2}{\left(-9 + i\right)\left(-9 - i\right)}\cos 3x$$
$$= \frac{2}{(82)}\cos 3x \qquad (3\text{-}51.5)$$
$$= \frac{1}{41}\cos 3x$$

Which is the particular solution (3-51.1)

(b) Alternative solution

Equation (3-49.8) allows us to write

$$y = \frac{2}{\left(D^4 + 1\right)}\cos 3x$$
$$= \frac{2}{\left(\left(-3^2\right)^2 + 1\right)}\cos 3x = \frac{1}{41}\cos 3x \qquad (3\text{-}51.6)$$

Example 59

Integrate the following o.d.e. using the Inverse D-Operator

$$\frac{d^4 y}{dx^4} + 9y = 5\sin x \qquad (3\text{-}52.1)$$

Solution

The D-Operator form is

$$\left(D^2 + 9\right)y = 5\sin x \qquad\qquad (3\text{-}52.2)$$

Equation (3-49.7) allows us to write

$$
\begin{aligned}
y &= \frac{5}{\left(D^2 + 9\right)}\sin x \\
&= \frac{5}{\left(-1 + 9\right)}\sin x \qquad\qquad (3\text{-}52.3) \\
&= \frac{5}{8}\sin x
\end{aligned}
$$

Which is the particular solution (3-52.1)

Example 60

Integrate the following o.d.e. using the Inverse D-Operator

$$\frac{d^2y}{dx^2} - 4\frac{dy}{dx} + 4y = x^2 e^{2x} \qquad\qquad (3\text{-}53.1)$$

Solution

The D-Operator form is

$$\left(D^2 - 4D + 4\right)y = x^2 e^{2x} \qquad\qquad (3\text{-}53.2)$$

Equation (3-49.9d) allows us to write

$$y = \frac{1}{\left(D^2 - 4D + 4\right)} x^2 e^{2x}$$

$$= \frac{1}{\left(D - 2\right)^2} x^2 e^{2x}$$

$$= e^{2x} \frac{1}{\left(D - 2 + 2\right)^2} x^2 \qquad (3\text{-}53.3)$$

$$= e^{2x} \frac{1}{D^2} x^2$$

Integrating twice, we get

$$y = e^{2x} \frac{1}{D^2} x^2$$

$$= e^{2x} \frac{1}{3D} x^3 \qquad (3\text{-}53.4)$$

$$= e^{2x} \frac{1}{12} x^3$$

Which is the particular solution (3-53.1)

Example 61

Integrate the following o.d.e. using the Inverse D-Operator

$$\frac{d^3 y}{dx^3} - 3 \frac{d^2 y}{dx^2} + 3 \frac{dy}{dx} - y = e^x \qquad (3\text{-}54.1)$$

Solution

The D-Operator form is

$$\left(D^3 - 3D^2 + 34D - 4\right)y = e^x \qquad (3\text{-}54.2)$$

(a) Equation (3-49.6) allows us to write

$$y = \frac{1}{\left(D^3 - 3D^2 + 3D - 1\right)}e^x$$

$$= \frac{1}{\left(D - 1\right)^3}e^x \qquad\qquad (3\text{-}54.3a)$$

$$= e^x \frac{1}{\left(-1+1\right)^3}$$

Hence, we face **the vanishing denominator, which negates the use of equation** (3-49.6).

(b) Alternative solution

Equation (3-49.9d) allows us to circumvent the vanishing denominator as follows

$$y = \frac{1}{\left(D - 1\right)^3}\left(e^x.1\right)$$

$$= e^x \frac{1}{\left(\left(D + 1\right) - 1\right)^3}1 \qquad\qquad (3\text{-}54.3b)$$

$$= e^x \frac{1}{D^3}1$$

Thus, instead of replacing D by 1, we added 1 to D. Integrating thrice, we get

$$y = e^x \frac{1}{D^2}x$$

$$= e^x \frac{1}{2D}x^2 \qquad\qquad (3\text{-}54.4)$$

$$= e^x \frac{1}{6}x^3$$

Example 62

Integrate the following o.d.e. using the Inverse D-Operator

$$\frac{d^3y}{dx^3} - y = \sin x \qquad\qquad (3\text{-}55.1)$$

Solution

The D-Operator form is

$$\left(D^3 - 1\right)y = \sin x \qquad (3\text{-}55.2)$$

(a) Euler's formula transformation

$$\left(D^3 - 1\right)y = e^{ix}$$
$$= \cos x + i \sin x \qquad (3\text{-}55.3)$$

Then, the imaginary part of the obtained solution comprises the required particular solution.

Equation (3-49.6) allows us to write

$$y = \frac{1}{\left(D^3 - 1\right)} e^{ix}$$
$$= e^{ix} \frac{1}{\left(i^3 - 1\right)}$$
$$= e^{ix} \frac{1}{\left(-i - 1\right)} = -e^{ix} \frac{1}{\left(i + 1\right)} \frac{\left(i - 1\right)}{\left(i - 1\right)} \qquad (3\text{-}55.4)$$
$$= \frac{\left(i - 1\right)}{2} (\cos x + i \sin x)$$
$$= \frac{1}{2} (i \cos x - \sin x - \cos x - i \sin x)$$

The imaginary part and sought solution is

$$y = \frac{1}{2}(\cos x - \sin x) \qquad (3\text{-}55.5)$$

(b) Alternative solution

We could also use equation (3-49.7) as follows

$$y = \frac{\sin x}{\left(D^3 - 1\right)} = \frac{\sin x}{\left(\left(D^2\right)D - 1\right)}$$
$$= \frac{\sin x}{\left(\left(-1\right)D - 1\right)} = -\frac{\sin x}{\left(D + 1\right)} \qquad (3\text{-}55.6)$$

Multiplying by (D-1)/(D-1), we get

$$y = -\frac{1}{(D+1)}\frac{(D-1)}{(D-1)}\sin x$$

$$= -\frac{(D-1)}{(D^2-1)}\sin x$$

$$= -\frac{(D-1)}{(-1-1)}\sin x = \frac{(D-1)}{2}\sin x$$

$$= \frac{(D\sin x - \sin x)}{2} = \frac{(\cos x - \sin x)}{2}$$

(3-55.7)

Example 63

Integrate the following o.d.e. using the Inverse D-Operator

$$\frac{d^2 y}{dx^2} + y = \cos x$$

(3-56.1)

Solution

The D-Operator form is

$$\left(D^2 + 1\right)y = \cos x$$

(3-56.2)

(a) Euler's formula transformation

$$\left(D^2 + 1\right)y = e^{ix}$$

$$= \cos x + i \sin x$$

(3-56.3)

Then, the real part of the obtained solution comprises the required particular solution.

Equation (3-49.6) allows us to write

$$y = \frac{1}{\left(D^2 + 1\right)}e^{ix}$$

$$= \frac{1}{\left(D^2 - i^2\right)}e^{ix}$$

$$= \frac{1}{(D+i)(D-i)}e^{ix}$$

(3-56.4)

Here, we have two problems to solve by equations (3-49.9d) and (3-49.6), as follows.

First, equation (3-49.6), can be used with the nonvanishing term (D+i). Thus

$$y = \frac{1}{(i+i)(D-i)}e^{ix} \tag{3-56.5}$$

Second, equation (3-49.9d), can be used with the nonvanishing term (D-i). Thus

$$
\begin{aligned}
y &= e^{ix}\frac{1}{(i+i)(D-i+i)}\\
&= e^{ix}\frac{1}{2i}\left(\frac{1}{D}1\right) = e^{ix}\frac{x}{2i}\\
&= (\cos x + i\sin x)\frac{x}{2i}\\
&= \frac{x}{2}(-i\cos x + \sin x)
\end{aligned} \tag{3-56.6}
$$

Thus, the real part and particular solution is

$$y = \frac{x}{2}\sin x \tag{3-56.7}$$

Example 64

Integrate the following o.d.e. using the Inverse D-Operator

$$\frac{d^4 y}{dx^4} - y = e^x \tag{3-57.1}$$

Solution

The D-Operator form is

$$(D^4 - 1)y = e^x \tag{3-57.2}$$

Thus,

$$
\begin{aligned}
y &= \frac{1}{(D^4 - 1)}e^x\\
&= \frac{1}{(D-1)(D+1)(D^2 +1)}e^x
\end{aligned} \tag{3-57.3}
$$

First, equation (3-49.6), can be used with the nonvanishing term (D+1). Thus

$$y = \frac{1}{(D-1)(1+1)(1+1)}e^x \tag{3-57.4}$$

135

Second, equation (3-49.9d), can be used with the nonvanishing term (D-1). Thus

$$y = e^x \frac{1}{4(D-1+1)}(1)$$

$$= e^x \frac{x}{4}$$

(3-57.5)

3-7. Inverse D-Operator on a polynomial

Given the polynomial of degree K as follows:

$$P_K(x) = \sum_{k=0}^{K} a_k x^k, \qquad \text{where } a_0 \neq 0$$

(3-58.1)

And the **D-Operator polynomial** of order n as follows:

$$F(D) = \sum_{j=0}^{n} b_j D^j$$

(3-58.2)

The Inverse of the D-Operator F(D) is obtained as follows

$$\frac{1}{F(D)} = Q_K(x) + \frac{R(D)}{F(D)}$$

(3-58.3a)

Or,

$$\sum_{j=0}^{n} b_j D^j \overline{\left| \left(\frac{\sum_{k=0}^{K} c_k D^k}{1} \right) \right.}$$

(3-58.3b)

Where, the **quotient** $Q_K(D)$ has the same order K of the degree K of the polynomial PK(x), (3-58.1) and is defined as

$$Q(D) = \sum_{j=0}^{K} c_j D^j$$

(3-58.4)

R(D) is the **remainder** of the division of the dividend unity by the divisor F(D), (3-58.3).

136

$$R(D) = \sum_{j=K+1}^{K+n} c_j D^j \qquad\qquad (3\text{-}58.5)$$

We note that the remainder $R(D)$ has greater order than the degree K of the polynomial $P_K(x)$, which implies that

$$R(D)P_K(x) = 0 \qquad\qquad (3\text{-}58.6)$$

Therefore, applying the polynomial of the Inverse D-Operator, (3-58.3) on the polynomial function, (3-58.1), gives

$$\frac{1}{F(D)} P_K(x) = \left(Q_K(x) + \frac{R(D)}{F(D)} \right) P_K(x) \qquad\qquad (3\text{-}58.7)$$
$$= Q_K(x)P_K(x)$$

We conclude that, the n-order polynomial of Inverse D-operator applies to the K-degree polynomial function by its differential quotient Q(D) of order K.

Example 65

Integrate the following o.d.e. using the Inverse D-Operator

$$\frac{d^2 y}{dx^2} + y = x^2 - x + 2 \qquad\qquad (3\text{-}59.1)$$

Solution

The D-Operator form is

$$\left(D^2 + 1 \right) y = x^2 - x + 2 \qquad\qquad (3\text{-}59.2)$$

137

$$\begin{array}{r|l} & \dfrac{1-D^2}{1} \\ \hline D^2+1 & \begin{array}{l} \dfrac{1}{-\left(D^2+1\right)} \\ \hline -D^2 \\ +\left(D^4+D^2\right) \\ \hline +D^4 \end{array} \end{array}$$

(3-59.3)

Therefore,

$$Q(D) = 1 - D^2$$
$$R(D) = +D^4$$

(3-59.4)

Thus, equation (3-58.7) allows us to write

$$\frac{1}{F(D)} P_K(x) = Q(D)P_K(x)$$
$$= \left(1 - D^2\right)\left(x^2 - x + 2\right)$$
$$= \left(x^2 - x + 2 - 2\right)$$
$$y = x^2 - x$$

(3-59.5)

Clearly,

$$R(D)P_K(x) = D^4\left(x^2 - x + 2\right)$$
$$= 0$$

(3-59.6)

Example 66

Integrate the following o.d.e. using the Inverse D-Operator

$$y'' + 2y' + 2y = x^2 e^{-x}$$

(3-60.1)

Solution

The D-Operator form is

$$\left(D^2 + 2D + 2\right)y = x^2 e^{-x}$$

(3-60.2)

The particular solution is

$$y = \frac{1}{\left(D^2 + 2D + 2\right)} x^2 e^{-x} \tag{3-60.3}$$

First, we use the shift property, equation (3-49.9d), to get the exponential term out of the above expression

$$y = e^{-x} \frac{1}{\left((D-1)^2 + 2(D-1) + 2\right)} x^2$$
$$= e^{-x} \frac{1}{\left(D^2 + 1\right)} x^2 \tag{3-60.4}$$

Then, we obtain the quotient

$$\tag{3-60.5}$$

Similar to (3-59.4),

$$Q(D) = 1 - D^2$$
$$R(D) = +D^4 \tag{3-60.6}$$

Thus, equation (3-58.7) allows us to write

$$y = e^{-x}\left(1 - D^2\right)x^2$$
$$= e^{-x}\left(x^2 - 2\right) \tag{3-60.7}$$

Example 67

Integrate the following o.d.e. using the Inverse D-Operator

$$y'' + y = x \cos x \tag{3-61.1}$$

Solution

The D-Operator form is

$$\left(D^2 + 1\right)y = x\cos x \tag{3-61.2}$$

Using Euler's formula, the particular solution is

$$y = \frac{1}{\left(D^2 + 1\right)}xe^{ix} \tag{3-61.3}$$

First, we use the shift property, equation (3-49.9d), to get the exponential term out of the above expression

$$\begin{aligned}
y &= e^{ix}\frac{1}{\left((D+i)^2 + 1\right)}x \\
&= e^{ix}\frac{1}{D(D + 2i)}x
\end{aligned} \tag{3-61.4}$$

Then, we obtain the quotient

$$\tag{3-61.5}$$

$$
\begin{array}{c}
\dfrac{1}{2i} - \dfrac{D}{(2i)^2} \\[2ex]
\hline
D + 2i \left|
\begin{array}{l}
\dfrac{1}{-\left(\dfrac{D}{2i} + 1\right)} \\[3ex]
\hline
-\dfrac{D}{2i} \\[2ex]
-\left(\dfrac{D^2}{4} + \dfrac{iD}{2}\right) \\[3ex]
\hline
-\dfrac{D^2}{4}
\end{array}
\right.
\end{array}
$$

Similar to (3-59.4),

$$\begin{aligned}
Q(D) &= \frac{1}{2i} + \frac{D}{4} \\
R(D) &= -\frac{D^2}{4}
\end{aligned} \tag{3-61.6}$$

140

Thus, equation (3-58.7) allows us to write

$$
\begin{aligned}
y &= e^{ix}\frac{1}{D(D+2i)}x \\
&= e^{ix}\frac{1}{D}Q(D)x \\
&= e^{ix}\frac{1}{D}\left(\frac{1}{2i}+\frac{D}{4}\right)x \\
&= e^{ix}\frac{1}{D}\left(\frac{x}{2i}+\frac{1}{4}\right)
\end{aligned}
\qquad (3\text{-}61.7)
$$

Performing the remaining integration, we get

$$
y = e^{ix}\left(\frac{x^2}{4i}+\frac{x}{4}\right)
\qquad (3\text{-}61.7)
$$

The particular solution of equation (3-61.1) is the real part of the above equation

$$
\begin{aligned}
y &= (\cos x + i\sin x)\left(\frac{x^2}{4i}+\frac{x}{4}\right) \\
&= \frac{x}{4}(x\sin + \cos x)
\end{aligned}
\qquad (3\text{-}61.8)
$$

3-8. Euler's Linear Equation

Euler's equation is written as

$$
\sum_{n=0}^{n} c_n (ax+b)^n D^n y = f(x)
\qquad (3\text{-}62.1)
$$

The following substitution transforms Euler's equation into a linear o.d.e. of constant coefficients:

$$
\begin{aligned}
ax+b &= e^t \\
t &= \ln(ax+b)
\end{aligned}
\qquad (3\text{-}62.2)
$$

Such that

$$
\frac{dt}{dx} = \frac{a}{ax+b}
\qquad (3\text{-}62.3)
$$

141

The D-Operator becomes

$$Dy = \frac{dy}{dt}\frac{dt}{dx} = \frac{a}{ax+b}\frac{dy}{dt}$$

(3-62.4)

Differentiating again, the second derivative, D^2, becomes

$$D^2 y = \frac{a}{ax+b}\frac{d^2 y}{dt^2}\frac{dt}{dx} - \left(\frac{a}{ax+b}\right)^2 \frac{dy}{dt}$$

$$= \left(\frac{a}{ax+b}\right)^2 \frac{d}{dt}\left(\frac{d}{dt}-1\right)y$$

(3-62.5)

Similarly,

$$D^n y = \left(\frac{a}{ax+b}\right)^n \frac{d}{dt}\left(\frac{d}{dt}-1\right)\left(\frac{d}{dt}-2\right)...\left(\frac{d}{dt}-n+1\right)y$$

(3-62.6)

Substituting from (3-62.6) in (3-62.1), we could eliminate the variable coefficients of the D-operator, get new structure of the form (3-62.6), hence a linear o.d.e. with constant coefficients.

Example 68

Integrate the following o.d.e. using the Inverse D-Operator

$$x^2 y'' - xy' + y = x(\ln x)^3$$

(3-63.1)

Solution

We note that x has the degree equal to the order of the differential derivative of y.

Therefore, we substitute by

$$x = e^t$$
$$t = \ln x$$

(3-63.2)

Such that

$$\frac{dt}{dx} = \frac{1}{x}$$

(3-63.3)

The D-Operator becomes

$$\frac{dy}{dx} = \frac{dy}{dt}\frac{dt}{dx} = \frac{1}{x}\frac{dy}{dt}$$

(3-63.4)

142

Differentiating again, the second derivative, D^2, becomes

$$\frac{d^2y}{dx^2} = \frac{1}{x^2}\frac{d^2y}{dt^2} - \frac{1}{x^2}\frac{dy}{dt} \qquad (3\text{-}63.5)$$

Substituting from (3-62.4) and (3-62.5) in (3-62.1), we get

$$\frac{d^2y}{dt^2} - 2\frac{dy}{dt} + y = e^t t^3 \qquad (3\text{-}63.6)$$

Denoting the new D-Operator by

$$D = \frac{d}{dt}$$

Equation (3-63.6) becomes

$$\left(D^2 - 2D + 1\right)y = e^t t^3 \qquad (3\text{-}63.7)$$

(a) Complementary function

Thus, the characteristic equation is

$$m^2 - 2m + 1 = 0 \qquad (3\text{-}63.8)$$

Which has two equal roots, $m = 1$, and hence the **Complementary function** of (3-63.7) is

$$y = \left(A + Bt\right)e^t \qquad (3\text{-}63.9)$$

(b) Particular Solution

From equation (3-63.7), we get

$$y = \frac{1}{\left(D - 1\right)^2}e^t t^3 \qquad (3\text{-}63.10)$$

Using the property of equation (3-49.9d) of shifting the D operator by the factor of the exponent of e, we get:

143

$$y = e^t \frac{1}{(D+1-1)^2} t^3$$
$$= e^t \frac{1}{D^2} t^3$$

(3-63.11)

Integrating twice, we get

$$y = e^t \frac{1}{20} t^5$$

(3-63.12)

(c) General Solution

From equations (3-63.9) and (3-63.12), we get

$$G.S. = C.F. + P.S.$$
$$= (A + Bt)e^t + e^t \frac{1}{20} t^5$$

(3-63.13)

Restoring the initial independent variable x by the substitution in equation (3-63.2), the general solution becomes

$$y. = x\left((A + B\ln x) + \frac{1}{20}(\ln x)^5 \right)$$

(3-63.14)

3-9. Solution of a system of spontaneous linear o.d.e.

The solution of a system of spontaneous linear differential equations (linear in all unknown functions and their derivatives) is achieved by obtaining a **single equation of higher order** such that all unknown functions **except one** are eliminated. The obtained single function is then integrated to yield on unknown function, which is then used to determine the remaining unknowns.

Example 69

Solve the two simultaneous o.d.e.'s

$$\frac{dy}{dt} = x$$
$$\frac{dx}{dt} = y$$

(3-64.1)

Solution

144

The D-Operator form is

$$Dy - x = 0$$
$$Dx - y = 0 \qquad (3\text{-}64.2)$$

First, differentiate one of the two equations, and then substitute into it by the other

$$D^2y - Dx = 0 \qquad (3\text{-}64.3)$$

Substitute by $Dx = y$, to get

$$D^2y - y = 0 \qquad (3\text{-}64.4)$$

The characteristic equation is

$$m^2 - 1 = 0 \qquad (3\text{-}64.5)$$

The two roots $m_1 = 1$, and $m_2 = -1$ lead to the solution

$$y = Ae^t + Be^{-t} \qquad (3\text{-}64.6)$$

Second, substitute by y, from (3-64.6) in (3-64.2) and integrate as follows

$$Dx = Ae^t + Be^{-t} \qquad (3\text{-}64.7)$$

Thus,

$$x = \frac{1}{D}Ae^t + \frac{1}{D}Be^{-t}$$
$$= Ae^t - Be^{-t} + C \qquad (3\text{-}64.7)$$

Example 70

Solve the two simultaneous o.d.e.'s

$$\frac{dy}{dt} = 2x - y$$
$$\frac{dx}{dt} = 3x - 2y \qquad (3\text{-}65.1)$$

Solution

The D-Operator form is

$$Dy = 2x - y$$
$$Dx = 3x - 2y$$
(3-65.2)

First, differentiate one of the two equations, and then substitute into it by the other

$$D^2y = 2Dx - Dy$$
(3-65.3)

We need to eliminate Dx from the above equation and replace it with a function in y only. Therefore, we start by eliminating x between the two equations in (3-65.2), to get

$$+(3Dy = 6x - 3y)$$
$$-(2Dx = 6x - 4y)$$
$$\overline{3Dy - 2Dx = y}$$
(3-65.4)

$$Dx = \frac{3Dy - y}{2}$$

Substituting by Dx, from the above equation, in (3-65.3), we get

$$D^2y = 2Dy - y \qquad\qquad s$$
(3-65.5)

The characteristic equation is

$$m^2 - 2m + 1 = 0$$
(3-65.5)

Thus, we have double and equal roots, leading to the solution

$$y = e^t(A + Bt)$$
(3-65.6)

Second, substitute by y, from (3-65.6) in (3-65.2) and integrate as follows

$$D\left[e^t(A + Bt)\right] = 2x - e^t(A + Bt)$$
(3-65.7)

Thus,

$$x = \frac{e^t(2A + B(2t + 1))}{2}$$
(3-65.8)

Example 71

Solve the two simultaneous o.d.e.'s

$$\frac{d^2y}{dt^2} = x$$

$$\frac{d^2x}{dt^2} = y$$

(3-66.1)

Solution

The D-Operator form is

$$D^2y = x$$
$$D^2x = y$$

(3-66.2)

First, differentiate one of the two equations twice

$$D^4y = D^2x$$

(3-66.3)

Substituting by D^2x, from (3-66.2) in the above equation, we get

$$D^4y = y$$

(3-66.4)

The characteristic equation is

$$m^4 - 1 = 0$$

Or

$$(m - i)(m + i)(m + 1)(m - 1) = 0$$

(3-66.5)

Thus, we have four distinct roots, leading to the solution

$$y = Ae^t + Be^{-t} + C\cos t + E\sin t$$

(3-65.6)

Second, differentiate y twice and substitute in (3-66.2) to obtain x, as follows

$$x = D^2y = Ae^t + Be^{-t} - C\cos t - E\sin t$$

(3-65.7)

3-10. Exercises on linear o.d.e. with constant coefficients

Solve the following o.d.e.

3-1. $\dfrac{d^2y}{dx^2} - 6\dfrac{dy}{dx} + 10y = 100$, for x = 0, y = 10, y'=5

                                 [Ans : $y = 5e^3 x \sin x + 10$]

3-2. $\dfrac{d^2y}{dx^2} + x = \sin t - \cos 2t$ [Ans : $x = A\cos t + B\sin t + \dfrac{1}{3}\cos 2t - \dfrac{1}{2}t\cos t$]

3-3. $x^2\dfrac{d^2y}{dx^2} - 4x\dfrac{dy}{dx} + 6y = 2$ [Ans : $y = Ax^2 + Bx^3 + \dfrac{1}{3}$]

3-4. $\dfrac{d^2y}{dx^2} + y = \cosh x$ [Ans : $y = A\sin x + B\cos x + \dfrac{1}{2}\cosh x$]

3-5. $\dfrac{d^2x}{dt^2} - 4\dfrac{dx}{dt} + 4x = e^t + e^{2t} + t$ [Ans : $x = e^{2t}(A + Bt) + \dfrac{1}{2}t^2e^{2t} + e^t + \dfrac{1}{4}$]

3-6. $\dfrac{d^4y}{dx^4} - 10y = x^2 - e^x$ $\left[\text{Ans}: \ x = Ae^{2t} + Be^{-2t} + C\cos 2x + E\sin 2x - \dfrac{x^2}{10} + \dfrac{e^x}{15}\right]$

3-7. $\dfrac{d^6x}{dt^6} - \dfrac{d^4x}{dt^4} = 1$ $\left[\text{Ans}: \ x = Ae^t + Be^{-t} + Ct^3 + Et^2 + Ft + G - \dfrac{t^4}{24}\right]$

3-8. $\dfrac{d^4x}{dt^4} - 2\dfrac{d^2x}{dt^2} + x = t^2 - 3$ $\left[\text{Ans}: \ x = e^t(A + Bt) + e^{-t}(E + Ft) + 1 + t^2\right]$

3-9. $\dfrac{d^2x}{dt^2} + 9x = t\sin 3t$ $\left[\text{Ans}: \ x = A\cos 3t + B\sin 3t - \dfrac{1}{12}t^2\cos 3t + \dfrac{1}{36}\sin 3t\right]$

3-10. $\dfrac{d^2y}{dx^2} + 2\dfrac{dy}{dx} + y = \sinh x$ $\left[\text{Ans}: \ y = e^{-x}(A + Bx - \dfrac{1}{4}x^2) + \dfrac{1}{8}e^x\right]$

3-11. $\dfrac{d^3y}{dx^3} - y = e^x$ $\left[\text{Ans}: \ y = Ae^x + e^{-\frac{1}{2}x}(B\cos\dfrac{\sqrt{3}}{2}x + C\sin\dfrac{\sqrt{3}}{2}x) + \dfrac{1}{3}xe^x\right]$

3-12. $\dfrac{d^2 y}{dx^2} - 2\dfrac{dy}{dx} + 2y = xe^x \cos x$

$$\left[\text{Ans}: \quad y = Ae^x (A\cos x + B\sin x) + \frac{1}{4} e^x B\cos x + \frac{1}{4} x^2 e^x \sin x \right]$$

3-13. $\dfrac{d^6 y}{dx^6} - 3\dfrac{d^5 y}{dx^5} + 3\dfrac{d^4 y}{dx^4} - \dfrac{d^3 y}{dx^3} = x$

$$\left[\text{Ans}: \quad y = A + Bx + Cx^2 + e^x (E + Fx + Gx^2) - \frac{1}{2} x^3 - \frac{1}{24} x^4 \right]$$

3-14. $\dfrac{d^4 y}{dt^4} + 2\dfrac{d^3 y}{dt^3} + x = \cos t$ $\quad \left[\text{Ans}: \quad x = (A + Bt)\cos t + (E + Ft)\sin t - \frac{1}{8} t^2 \cos t \right]$

3-15. $(1+x)^2 \dfrac{d^2 y}{dx^2} + (1+x)\dfrac{dy}{dx} + y = 2\cos \ln(1+x)$

$$\left[\text{Ans}: \quad y = A\cos \ln(1+x) + B\sin \text{lin}(1+x) + \ln(1+x)\sin \ln(1+x) \right]$$

3-16. $\dfrac{d^4 x}{dt^4} + x = t^3$

$$\left[\text{Ans}: \quad x = e^{t/\sqrt{2}} \left(A\cos \frac{t}{\sqrt{2}} + B\sin \frac{t}{\sqrt{2}} \right) + e^{-t/\sqrt{2}} \left(E\cos \frac{t}{\sqrt{2}} + F\sin \frac{t}{\sqrt{2}} \right) + t^3 \right]$$

3-17. $\dfrac{d^2 x}{dt^2} + 10\dfrac{dx}{dt} + 25x = 2^t + e^{-5t}$ $\quad \left[\text{Ans}: \quad x =)A + Bt)e^{-5t} + \frac{2t}{(5 + \ln x)^2} + \frac{1}{6} t^3 e^{-5t} \right]$

3-18. $\dfrac{d^6 y}{dx^6} - y = e^{2x}$

$$\left[\begin{aligned} \text{Ans}: y &= Ae^x + Be^{-x} + e^{\frac{x}{2}} \left(E\cos \frac{\sqrt{3}}{2} x + F\sin \frac{\sqrt{3}}{2} x \right) \\ &\quad + e^{-\frac{x}{2}} \left(G\cos \frac{\sqrt{3}}{2} x + B\sin \frac{\sqrt{3}}{2} x \right) + \frac{1}{63} e^{2x} \end{aligned} \right]$$

3-19.
$$\frac{d^6y}{dx^6} + 2\frac{d^4y}{dx^4} + \frac{d^2y}{dx^2} = x + e^x$$

$$\left[\text{Ans}: \quad y = (A + Bx)\cos x + (E + Fx)\sin x + G + Hx + \frac{1}{6}x^3 + \frac{1}{4}e^x \right]$$

3-20.
$$\frac{d^2y}{dx^2} + y = \sin 3x \cos x$$

$$\left[\text{Ans}: \quad y = A\cos x + B\sin x - \frac{1}{6}\sin 2x - \frac{1}{30}\sin 4x \right]$$

Solve the following system of simultaneous linear o.d.e.

3-21.
$$\frac{dx}{dt} = y, \qquad \frac{dy}{dt} = -x, \qquad x(0) = 0, \qquad y(0) = 1$$

$$\left[\text{Ans}: \quad x = \sin t, \qquad y = \cos t \right]$$

3-22.
$$\frac{d^2x}{dt^2} = y, \qquad \frac{d^2y}{dt^2} = x, \qquad x(0) = x, \qquad \frac{dx}{dt}(0) = 2, \qquad y(0) = 2, \qquad \frac{dy}{dt}(0) = 2$$

$$\left(\text{Ans}: \quad x = 2e^t, \quad x = 2e^t \right)$$

3-23.
$$\frac{dx}{dt} + 5x + y = e^t, \qquad \frac{dy}{dt} - x - 3y = e^{2t}$$

$$\left(\begin{array}{l} \text{Ans}: \quad x = Ae^{(-1+\sqrt{15})t} + Be^{(-1-\sqrt{15})t} + \frac{2}{11}e^t + \frac{1}{6}e^{2t} \\[2mm] \quad\quad\quad y = e^t - \frac{dx}{dt} - 5x \end{array} \right)$$

3-24.
$$\frac{dx}{dt} = y, \qquad \frac{dy}{dt} = z, \qquad \frac{dz}{dt} = x$$

$$\left(\text{Ans}: \quad x = Ae^t + e^{\frac{-1}{2}t}\left(B\cos\frac{\sqrt{3}}{2}t + C\sin\frac{\sqrt{3}}{2}t \right), \qquad y = \frac{dx}{dt}, \qquad z = \frac{d^2x}{dt^2} \right)$$

3-25.
$$\frac{dx}{dt} = y, \quad \frac{dy}{dt} = \frac{y^2}{x}$$
$$\left(\text{Ans}: \ x = Ae^{Bt}, \quad y = ABe^{Bt} \right)$$

3-26.
$$\frac{dx}{dt} + \frac{dy}{dt} = -x + y + 3, \quad \frac{dx}{dt} - \frac{dy}{dt} = x + y - 3$$
$$\left(\text{Ans}: \ x = A\cos t + B\sin t + 3, \quad y = -A\sin t + B\cos t \right)$$

3-27.
$$\frac{dx}{dt} = -x + y + z, \quad \frac{dy}{dt} = x - y + z, \quad \frac{dx}{dt} = x + y - z$$
$$\left(\text{Ans}: \ x = Ae^{t} + Be^{-2t}, \quad y = Ae^{t} + Ce^{-2t}, \quad z = Ae^{t} - (B + C)e^{-2t} \right)$$

3-28.
$$t\frac{dx}{dt} + y = 0, \quad t\frac{dy}{dt} + x = 0$$
$$\left(\text{Ans}: \ x = At + \frac{B}{t}, \quad y = -At + \frac{B}{t} \right)$$

3-29.
$$\frac{dx}{dt} + y = \cos t, \quad \frac{dy}{dt} + x = \sin t \left(\text{Ans}: \ x = Ae^{t} + Be^{-t}, \quad y = -Ae^{t} + Be^{-t} \right)$$

3-30.
$$\frac{dx}{dt} + 3x - y = 0, \quad \frac{dy}{dt} - 8 + y = 0, \quad x(0) = 1,, \quad y(0) = 4$$
$$\left(\text{Ans}: \ x = e^{t} \quad y = 4e^{t} \right)$$

3-31.
$$\ddot{x}(t) = ax - y, \quad \ddot{y}(t) = x + ay, \quad \text{a is a constant}$$
$$\left[\text{Ans}: \ x = e^{at}(A\cos t + B\sin t); \quad y = e^{at}(A\sin t - B\cos t) \right]$$

3-32. $\dfrac{dx}{dt} + 3x + 4y = 0, \qquad \dfrac{dy}{dt} + 2x + 5y = 0$

$\left(\text{Ans}: \quad x = 2Ae^{-t} + Be^{-7t} + \qquad y = -Ae^{-t} + Be^{-7t} \right)$

3-33. $\dfrac{dx}{dt} = -5x - 2y, \qquad \dfrac{dy}{dt} = -7y$

$\left(\text{Ans}: \quad x = e^{-6t}\left(2A\cos t + 2B\sin t\right), \quad y = e^{-6t}\left((A - B)\cos t + (A + B)\sin t\right) \right)$

3-34. $\dfrac{dx}{dt} = y - z, \qquad \dfrac{dy}{dt} = x + y, \qquad \dfrac{dz}{dt} = x + z$

$\left(\text{Ans}: \quad x = Ae^{t} + B, \quad y = e^{t}\left(At + C\right) - B, \quad z = e^{t}\left(At + C - A\right) - B \right)$

3-35. $\dfrac{dx}{dt} - y + z = 0, \qquad \dfrac{dy}{dt} - x - y = 1, \qquad \dfrac{dz}{dt} - x - z = 1$

$\left(\text{Ans}: \quad x = Ae^{t} + B, \quad y = e^{t}\left(At + C\right) - t - 1 - B, \quad z = y - Ae^{t} \right)$

3-36. $\dfrac{dy}{dx} = y + 4z, \qquad \dfrac{dz}{dx} + y + 3x = 0$

$\left(\text{Ans}: \quad y = e^{-x}\left(A\cos x + B\sin x\right), \quad z = \dfrac{1}{5}e^{-x}\left[(B - 2A)\cos x - (A + 2B)\sin x\right] \right)$

3-37. $4\dfrac{dx}{dt} - \dfrac{dy}{dt} + 3x = \sin t, \qquad \dfrac{dx}{dt} + y = \cos t$

$\left(\text{Ans}: \quad x = Ae^{-x} + Be^{-3x} \quad y = Ae^{-x} + 3Be^{-3x} + \cos t \right)$

Solve the following o.d.e.

3-38. $\dfrac{d^2y}{dx^2} - 2\dfrac{dy}{dx} + 3y = e^{-x}\cos x$

$$\left[\text{Ans}: \ y = e^x\left(A\cos\sqrt{2}x + B\sin\sqrt{2}x\right) + \frac{1}{41}e^{-x}\left(5\cos x - 4\sin x\right)\right]$$

3-39. $\dfrac{d^4y}{dx^4} - a^4y = 5a^4e^{ax}\sin ax$

$$\left[\text{Ans}: \ y = e^{ax}(A - \sin ax) + Be^{-ax} + C\cos ax + E\sin ax\right]$$

3-40. $\dfrac{d^2y}{dx^2} + \dfrac{dy}{dx} - 6y = xe^{2x}$

$$\left[\text{Ans}: \ y = Ae^{2x} + Be^{-3x} + x\left(\frac{1}{10} - \frac{1}{25}\right)e^{2x}\right]$$

3-41. $\dfrac{d^2y}{dx^2} - 2\dfrac{dy}{dx} + y = \sin + \sinh x$

$$\left[\text{Ans}: \ y = (A + Bx)e^x + \frac{1}{2}\cos x + \frac{x^2}{4}Be^x - \frac{1}{8}e^{-x}\right]$$

3-42. $\dfrac{d^2y}{dx^2} - 2\dfrac{dy}{dx} + 5y = e^x\cos 2x$

$$\left[\text{Ans}: \ y = (A\cos 2x + B\sin 2x)e^x + \frac{x}{4}e^x\sin 2x\right]$$

3-43. $\dfrac{d^2y}{dx^2} - 7\dfrac{dy}{dx} + 12y = -e^{4x}$

$$\left[\text{Ans}: \ y = Ae^{3x} + Be^{4x} - xe^{4x}\right]$$

3-44. $\dfrac{d^2y}{dx^2} - 2\dfrac{dy}{dx} - 8y = e^x - 8\cos 2x$

$$\left[\text{Ans}: \ y = Ae^{-2x} + Be^{4x} - \frac{1}{9}e^x + \frac{1}{5}(3\cos 2x + \sin 2x)\right]$$

3-45. $\dfrac{d^2y}{dx^2} - 2\dfrac{dy}{dx} + 8y = e^x + \sin 3x$

$$\left[\text{Ans}: \ y = (A\cos 3x + B\sin 3x)e^x + \dfrac{1}{37}(\sin 3x + 6\cos 3x) + \dfrac{1}{9}e^x \right]$$

3-46. $\dfrac{d^2y}{dx^2} - 3\dfrac{dy}{dx} = x + \cos x$

$$\left[\text{Ans}: \ y = A + Be^{3x} - \dfrac{1}{10}(\cos x + 3\sin x) - \dfrac{1}{6}x^2 - \dfrac{1}{9}x \right]$$

3-47. $\dfrac{d^2y}{dx^2} - 4y = e^{2x}\sin 2x$

$$\left[\text{Ans}: \ y = Ae^{-2x} + Be^{2x} - \dfrac{1}{10}e^{2x}(\sin 2x + 2\cos 2x) \right]$$

3-48. $\dfrac{d^2y}{dx^2} - 2\dfrac{dy}{dx} + 2y = 4e^x\sin x$ $\qquad \left[\text{Ans}: \ y = e^x(A\cos x + B\sin x - 2x\cos x) \right]$

3-49. $\dfrac{d^2y}{dx^2} = xe^x + y$

$$\left[\text{Ans}: \ y = Ae^x + Be^{-x} + \dfrac{1}{4}(x^2 - x)e^x \right]$$

3-50. $\dfrac{d^2y}{dx^2} + 9y = 2x\sin x + xe^{3x}$

$$\left[\text{Ans}: \ y = A\cos 3x + B\sin 3x + \dfrac{1}{4}x\sin x - \dfrac{1}{16}\cos x + \dfrac{e^{3x}}{54}(3x - 1) \right]$$

3-51. $\dfrac{d^2y}{dx^2} + 2\dfrac{dy}{dx} - 3y = 2xe^{-3x} + (x+1)e^x$

$$\left[\text{Ans}: \ y = Ae^{-3x} + Be^x - \dfrac{1}{8}(2x^2 + x)e^{-3x} + \dfrac{1}{16}(2x^2 + 3x)e^x \right]$$

3-52. $(3x + 2)\dfrac{d^2y}{dx^2} + 7\dfrac{dy}{dx} = 0$ $\qquad\qquad\qquad \left[\text{Ans}: \ y = A + B(3x + 2)^{-4/3} \right]$

3-53. $x^2 \dfrac{d^2y}{dx^2} - 4x\dfrac{dy}{dx} + 6y = x$ $\left[\text{Ans}: \ y = Ax^3 + Bx^2 + \dfrac{1}{2}x\right]$

3-54. $(1+x)\dfrac{d^2y}{dx^2} - 3\dfrac{dy}{dx} + \dfrac{4}{1+x}y = (1+x)^2$

$\left[\text{Ans}: \ y = (1+x)^2[A + B\ln(1+x)] + (1+x)^3\right]$

3-55. $\dfrac{d^2y}{dx^2} - \dfrac{1}{x}\dfrac{dy}{dx} + \dfrac{1}{x^2}y = \dfrac{2}{x}$ $\left[\text{Ans}: \ y = x[\ln x + (\ln x)^2]\right]$

CHAPTER 4

SECOND-ORDER
LINEAR DIFFERENTIAL EQUATIONS
WITH VARIABLE COEFFICIENTS

The second-order o.d.e. with **variable coefficients** takes the form

$$\frac{d^2y}{dx^2} + P(x)\frac{dy}{dx} + Q(x)y = f(x) \tag{4-1}$$

The variable coefficients $P(x)$ and $Q(x)$ are functions in x alone, rendering the equation **linear** in the **unknown dependent function y and its derivatives**.

4-1. Reduction of the order of homogeneous linear o.d.e. by substitution by a known particular solution

We have discussed this method in equations (3-20) by the substitution

$$y = y_1 \int u\,dx \tag{4-2.1}$$

Where, the unknown function u is defined as

$$\frac{dz}{dx} = u, \qquad z = \int u\,dx \tag{4-2.3}$$

And

$$y = y_1 z \tag{4-2.4}$$

Where, y_1 is a known particular solution of the homogenous part of the equation.

Example 72

Solve the o.d.e.

$$x^2 \frac{d^2y}{dx^2} - 3x \frac{dy}{dx} + 3y = 2x^3 \qquad (4\text{-}3.1)$$

Show that the order of the equation could be reduced by unity by substituting by a known particular solution.

Solution

(a) The **homogeneous part** of equation (4-3.1) is

$$x^2 \frac{d^2y}{dx^2} - 3x \frac{dy}{dx} + 3y = 0 \qquad (4\text{-}3.2)$$

The particular solution x^3 clearly satisfies the above equation as follows

$$x^2 \frac{d^2}{dx^2}(x^3) - 3x \frac{d}{dx}(x^3) + 3(x^3) = 0 \qquad (4\text{-}3.3)$$
$$x^2(6x) - 3x(3x^2) + 3(x^3) = 0$$

(b) Substituting by a product of particular solution and unknown function

Substitute in equation (4-3.2) by x^3z and its derivatives as follows:

$$y = x^3z \qquad (4\text{-}3.4)$$

Where, z is unknown function to be determined as follows.

Differentiate twice we get

$$\frac{dy}{dx} = x^3 \frac{dz}{dx} + 3x^2 z$$
$$\frac{d^2y}{dx^2} = x^3 \frac{d^2z}{d^2x} + 6x^2 \frac{dz}{dx} + 6xz \qquad (4\text{-}3.5)$$

Substituting from (4-3.4) and (4-3.5) in equation (4-3.2), we get

$$x^2 \left(x^3 \frac{d^2z}{d^2x} + 6x^2 \frac{dz}{dx} + 6xz \right) - 3x \left(x^3 \frac{dz}{dx} + 3x^2 z \right) + 3\left(x^3 z \right) = 2x^3$$

Arranging, we get

157

$$x^5 \frac{d^2z}{d^2x} + 6x^4 \frac{dz}{dx} + 6x^3z - 3x^4 \frac{dz}{dx} - 9x^3z + 3x^3z = 2x^3$$

$$x^5 \frac{d^2z}{d^2x} + 3x^4 \frac{dz}{dx} = 2x^3$$

Or,

$$x \frac{d^2z}{d^2x} + 3 \frac{dz}{dx} = \frac{2}{x^2} \qquad\qquad (4\text{-}3.6)$$

Thus, we have the transformed second-order equation in y and its derivatives, equation (4-3.2), onto the first-order differential equation in z' and its derivative z".

This can be farther simplified by denoting

$$z' = u \text{ and } z'' = u', \qquad\qquad (4\text{-}3.7a)$$

we get

$$\frac{du}{dx} + \frac{3u}{x} = \frac{2}{x^2} \qquad\qquad (4\text{-}3.7b)$$

(c) Reduced second-order non-homogeneous o.d.e.

$$\frac{du}{dx} + \frac{3u}{x} = \frac{2}{x^2}$$

$$\frac{du}{dx} + P(x)u = Q(x) \qquad\qquad (4\text{-}3.8)$$

The **integrating factor**, equations (1-17.4), is obtained as follows

$$\frac{d\lambda}{\lambda} = P(x)dx$$

$$= \frac{3}{x}dx \qquad\qquad (4\text{-}3.9a)$$

$$\ln \lambda = 3\ln x$$

$$\lambda = x^3 \qquad\qquad (4\text{-}3.9b)$$

$$u\lambda = \int \lambda Q dx + C$$

$$= \int x^3\left(\frac{2}{x^2}\right)dx + C \qquad (4\text{-}3.10)$$

$$= x^2 + C$$

$$u = \frac{1}{x} + \frac{C}{x^3} \qquad (4\text{-}3.11)$$

(d) Reverting from u to z

From (4-3.7a), we obtain z by integrating u, as follows

$$z = \int u dx$$

$$= \int \left(\frac{1}{x} + \frac{C}{x^3}\right)dx \qquad (4\text{-}3.12)$$

$$= \ln x + \frac{A}{x^2} + B$$

(e) Reverting from z to y

From (4-3.4), we get

$$y = x^3 z = x^3\left(\ln x + \frac{A}{x^2} + B\right) \qquad (4\text{-}3.13)$$

$$= Ax + x^3\left(\ln x + B\right)$$

Example 73

Solve the o.d.e.

$$\frac{d^2 y}{dx^2} + \frac{1}{x}\frac{dy}{dx} - \frac{1}{x^2}y = 5x^2 \qquad (4\text{-}4.1)$$

Show that the order of the equation could be reduced by unity by substituting by a known particular solution.

Solution

(a) The **homogeneous part** of equation (4-4.1) is

$$\frac{d^2y}{dx^2} + \frac{1}{x}\frac{dy}{dx} - \frac{1}{x^2}y = 0 \tag{4-4.2}$$

The particular solution x clearly satisfies the above equation as follows

$$\frac{d^2y}{dx^2} + \frac{1}{x}\frac{dy}{dx} - \frac{1}{x^2}y = 0$$
$$0 + \frac{1}{x} - \frac{1}{x^2}x = 0 \tag{4-4.3}$$

(b) Substituting by a product of particular solution and unknown function

Substitute in equation (4-4.2) by xz and its derivatives as follows:

$$y = xz \tag{4-4.4}$$

Where, z is unknown function to be determined as follows.

Differentiate twice we get

$$\frac{dy}{dx} = x\frac{dz}{dx} + z$$
$$\frac{d^2y}{dx^2} = x\frac{d^2z}{d^2x} + 2\frac{dz}{dx} \tag{4-4.5}$$

Substituting from (4-4.4) and (4-4.5) in equation (4-4.2), we get

$$\left(x\frac{d^2z}{dx^2} + 2\frac{dz}{dx}\right) + \frac{1}{x}\left(x\frac{dz}{dx} + z\right) - \frac{1}{x^2}xz = 5x^2$$
$$x\frac{d^2z}{dx^2} + 3\frac{dz}{dx} = 5x^2$$

Or,

$$\frac{d^2z}{d^2x} + \frac{3}{x}\frac{dz}{dx} = 5x \tag{4-4.6}$$

Thus, we have the transformed second-order equation in y and its derivatives, equation (4-4.2), onto the first-order differential equation in z' and its derivative z''.

This can be farther simplified by denoting

$$z' = u \text{ and } z'' = u',$$ (4-4.7a)

we get

$$\frac{du}{dx} + \frac{3u}{x} = 5x$$ (4-4.7b)

(c) Reduced second-order non-homogeneous o.d.e.

$$\frac{du}{dx} + \frac{3u}{x} = 5x$$
$$\frac{du}{dx} + P(x)u = Q(x)$$ (4-4.8)

The **integrating factor**, equations (1-17.4), is obtained as follows

$$\frac{d\lambda}{\lambda} = P(x)dx$$ (4-4.9a)
$$= \frac{3}{x}dx$$
$$\ln \lambda = 3\ln x$$ (4-4.9b)
$$\lambda = x^3$$

$$u\lambda = \int \lambda Q dx + C$$
$$= \int x^3(5x)dx + C$$ (4-4.10)
$$= x^5 + C$$

$$u = x^2 + \frac{C}{x^3}$$ (4-4.11)

(d) Reverting from u to z

From (4-4.7a), we obtain z by integrating u, as follows

$$z = \int u\,dx$$

$$= \int \left(x^2 + \frac{C}{x^3} \right) dx \qquad (4\text{-}4.12)$$

$$= \frac{x^3}{3} + \frac{A}{x^2} + B$$

(e) Reverting from z to y

From (4-4.4), we get

$$y = xz = x\left(\frac{x^3}{3} + \frac{A}{x^2} + B \right)$$

$$= \frac{x^4}{3} + \frac{A}{x} + Bx \qquad (4\text{-}4.13)$$

4-2. Solution of linear second-order o.d.e. by substitution by a product of two unknown arbitrary functions

We have discussed the method of reduction of linear o.d.e. onto simpler or **normal form** that could be easily integrated in equations (1-19).

Here, we apply the same method to second-order linear o.d.e. as follows.

Given the differential equation

$$\frac{d^2y}{dx^2} + P(x)\frac{dy}{dx} + Q(x)y = F(x) \qquad (4\text{-}5.1)$$

With a proposed solution of the form

$$y = f(x)g(x) \qquad (4\text{-}5.2)$$

Differentiating (4-5.2), we get

$$\frac{dy}{dx} = f(x)\frac{dg(x)}{dx} + g(x)\frac{df(x)}{dx} \qquad (4\text{-}5.3a)$$

$$\frac{d^2y}{dx^2} = f(x)\frac{d^2g(x)}{dx^2} + 2\frac{dg(x)}{dx}\frac{df(x)}{dx} + g(x)\frac{d^2f(x)}{dx^2} \qquad (4\text{-}5.3a)$$

162

Substituting from (4-5.3) in (4-5.1), we get

$$\left(fg''+2g'f'+gf''\right)+ P(x)\left(fg'+gf'\right)+ Q(x)gf = F(x) \tag{4-5.4}$$

Where we dropped the bracketed arguments to simplify notation.

We could lump the terms in equation (4-5.4) in such manner that simplifies the utility of the separation of the two variables, f and g, as follows

$$fg'' + \left[2f' + P(x)f\right]g' + \left[f'' + P(x)f' + Q(x)f\right]g = F(x) \tag{4-5.5a}$$

Equation (4-5.5) represents the main formula on which the present method is founded. **Equate the smaller bracketed term to zero,** which determines one of the two unknown functions, g and f.

Also, equation (4-5.5) can be farther simplified to take the normal form of a second-order o.d.e. as follows:

$$g'' + \left[\frac{2f' + P(x)f}{f}\right] g' + \left[\frac{f'' + P(x)f' + Q(x)f}{f}\right] g = \frac{F(x)}{f} \tag{4-5.5b}$$

Therefore,

$$2\frac{df}{dx} + Pf = 0 \tag{4-5.6}$$

Equation (4-5.6) determines f(x) by integration and simplifies equation (4-5.5) to

$$g'' + \left[\frac{f'' + P(x)f' + Q(x)f}{f}\right] g = \frac{F(x)}{f} \tag{4-5.7a}$$

Equation (4-7.7a) is known as the **normal form** of the second-order o.d.e. and can be farther simplified as follows:

$$g''(x) + \Psi(x)g(x) = X(x) \tag{4-5.7b}$$

Where,

$$\Psi(x) = \frac{f'' + P(x)f' + Q(x)f}{f}$$

$$X(x) = \frac{F(x)}{f} \tag{4-5.7c}$$

Equation (4-5.7b) represents the **normal form** to which all differential equations of the form (4-5.1) can be reduced.

163

Also, the term $\Psi(x)$ defines **Euler's equation** (3-62.1) when

$$\Psi(x) = \left(\frac{a}{ax+b}\right)^2$$

Integrating (4-5.6), we get

$$\int \frac{df}{f} = -\frac{1}{2}\int P dx + \ln C$$

$$f(x) = Ce^{-\frac{1}{2}\int P dx}$$

(4-5.8)

Hence, we determine one of the two functions f and g, of (4-5.2).

Then substitute by f from (4-5.8) into equation (4-5.7), and integrate to obtain g(x), and finally, the final solution of (4-5.1).

Example 74

Solve the differential equation by the reduction method

$$\frac{d^2y}{dx^2} - \frac{1}{x}\frac{dy}{dx} + \frac{y}{x^2} = \sqrt{x}\ln x$$

(4-6.1)

Solution

(a) Substitution by y(x) = g(x)f(x)

Substitute in equation (4-6.1) by $y = f(x)\, g(x)$, equation (4-5.2), and its derivatives, equation (4-5.3) to get

$$\left(fg''+2g'f'+gf''\right) - \frac{1}{x}\left(fg'+gf'\right) + \frac{1}{x^2}gf = \sqrt{x}\ln x$$

(4-6.2)

i.e.,

$$fg'' + \left(2f' - \frac{1}{x}f\right)g' + \left(f'' + \frac{1}{x^2}f - \frac{1}{x}f'\right)g = \sqrt{x}\ln x$$

(4-6.3)

Equate the left bracketed term to zero

$$2\frac{df}{dx} - \frac{f}{x} = 0$$

(4-6.4)

Arranging and integrating, we get

$$\int \frac{df}{f} = \int \frac{dx}{2x}$$

$$\ln f = \ln \sqrt{x} + \ln C$$

i.e.,

$$f = C\sqrt{x} \qquad (4\text{-}6.5)$$

Substitute by g from (4-6.5) into equation (4-6.3), and noting the vanishing bracket, we get

$$C\sqrt{x}g'' + C\left(-\frac{1}{4}\left(\frac{1}{x}\right)^{\frac{3}{2}} + \frac{1}{x^2}\sqrt{x} - \frac{1}{x}\frac{1}{2\sqrt{x}} \right)g = \sqrt{x}\ln x \qquad (4\text{-}6.6)$$

Or,

$$Cg'' + C\frac{1}{4x^2}g = \ln x \qquad (4\text{-}6.7)$$

Since C' is an arbitrary constant.

Equation (4-6.7) is the Euler's equation and is integrated by substituting by

$$
\begin{aligned}
&x = e^t \\
&t = \ln x \\
&\frac{dt}{dx} = \frac{1}{x} \\
&\frac{dg}{dx} = \frac{dg}{dt}\frac{dt}{dx} = \frac{1}{x}\frac{dg}{dt} \\
&\frac{d^2g}{dx^2} = \frac{1}{x^2}\frac{d^2g}{dt^2} - \frac{1}{x^2}\frac{dg}{dt}
\end{aligned}
\qquad (4\text{-}6.8)
$$

Euler's equation (4-6.7) becomes

$$C\left(\frac{1}{x^2}\frac{d^2g}{dt^2} - \frac{1}{x^2}\frac{dg}{dt} \right) + C\frac{1}{4x^2}g = \ln x \qquad (4\text{-}6.9)$$

$$\frac{d^2g}{dt^2} - \frac{dg}{dt} + \frac{1}{4}g = Ax^2\ln x$$

$$\left(D^2 - D + \frac{1}{4} \right)g = A\left(te^{2t}\right) \qquad (4\text{-}6.10)$$

(b) Complementary Function

165

The characteristic equation

$$\left(m - \frac{1}{2}\right)^2 = 0$$ (4-6.11)

Since we have two real and equal roots, then complementary function is

$$g(t) = \left(A + Bt\right) e^{\frac{1}{2}}$$ (4-6.12)

(c) Particular Solution

The particular solution of equation (4-6.10) is obtained as follows

$$
\begin{aligned}
g(t) &= \frac{1}{\left(D^2 - D + \frac{1}{4}\right)} A\left(te^{2t}\right) \\
&= e^{2t} \frac{A}{\left(D - \frac{1}{2} + 2\right)^2}(t) \\
&= e^{2t} \frac{A}{\left(D + \frac{3}{2}\right)^2}(t) \\
&= e^{2t} \frac{A}{\left(D^2 + 3D + \frac{9}{4}\right)}(t)
\end{aligned}
$$ (4-6.13)

The quotient of division of by the F(D) is obtained as follows

$$\begin{array}{r} \dfrac{4}{9}-\dfrac{4.4}{3.9}D \\ \hline 1 \end{array}$$

$$-\dfrac{4}{9}D^2-\dfrac{4}{3}D-1$$

$$D^2+3D+\dfrac{9}{4}\ \overline{\left)\ -\dfrac{4}{9}D^2-\dfrac{4}{3}D\right.}$$

$$+\dfrac{16}{27}D^3+\dfrac{16}{9}D^2+\dfrac{4}{3}D$$

$$+\dfrac{16}{27}D^3+\dfrac{12}{9}D^2$$

(4-6.14)

Substituting by the quotient $\dfrac{4}{9}-\dfrac{16}{27}D$, we get

$$g(t)=e^{2t}\left(\dfrac{4}{9}-\dfrac{16}{27}D\right)(t)$$

$$=\dfrac{4}{27}e^{2t}(3t-4)$$

(4-6.15)

(g) General Solution

From equations (4-6.12) and (4-6.15), we get

$$g(t)=\text{C.F.}+\text{P.S.}$$

$$=\left(A+Bt\right)e^{\frac{1}{2}}+\dfrac{4}{27}e^{2t}(3t-4)$$

(4-6.16)

Retrieving the x variable from equation (4-6.8), we get

$$g(x)=\left(A+Bt\right)\sqrt{x}+\dfrac{4}{27}x^2(3\ln x-4)$$

(4-6.17)

(e) Final Solution

The final solution is obtained from (4-6.5) and (4-6.17), the final general solution is

$$y(x)=f(x)g(x)=\left[\left(A+Bt\right)\sqrt{x}+\dfrac{4}{27}x^2(3\ln x-4)\right]\sqrt{x}$$

(4-6.18)

Example 75

Solve the differential equation by the reduction method

$$\frac{d^2y}{dx^2} - 2x\frac{dy}{dx} + (x^2 + 3)y = 3e^{\frac{1}{2}x^2}\sin x \qquad (4\text{-}7.1)$$

Solution

(a) Substitution by y(x) = g(x)f(x)

Substitute in equation (4-7.1) by $y = f(x)g(x)$, equation (4-5.2), and its derivatives, equation (4-5.3) to get

$$(fg''+2g'f'+gf'') - 2x(fg'+gf') + (x^2+3)gf = 3e^{\frac{1}{2}x^2}\sin x \qquad (4\text{-}7.2)$$

i.e.,

$$fg''+(2f'-2xf)g'+(f''-2xf'+(x^2+3)f)g = 3e^{\frac{1}{2}x^2}\sin x \qquad (4\text{-}7.3)$$

Equate the left bracketed term to zero

$$2\frac{df}{dx} - 2xf = 0 \qquad (4\text{-}7.4)$$

Arranging and integrating, we get

$$\int \frac{df}{f} = \int x\,dx$$

$$\ln f = \frac{x^2}{2} + \ln C$$

i.e.,

$$f = Ce^{\frac{x^2}{2}} \qquad (4\text{-}7.5)$$

Substitute by g from (4-7.5) into equation (4-7.3), and noting the vanishing bracket, we get

$$Ce^{\frac{x^2}{2}}g''+C\left(x^2e^{\frac{x^2}{2}} +e^{\frac{x^2}{2}} - 2x^2e^{\frac{x^2}{2}} +(x^2+3)e^{\frac{x^2}{2}}\right)g = 3e^{\frac{1}{2}x^2}\sin x \qquad (4\text{-}7.6)$$

Or,

$$g''+4g = 3A\sin x \qquad (4\text{-}7.7)$$

Since A is an arbitrary constant.

(b) Complementary Function

The characteristic equation

$$m^2 + 4 = 0 \qquad (4\text{-}7.8)$$

Since we have the two distinct imaginary roots $\pm 2i$, then complementary function is

$$g(t) = A\cos 2x + B\sin 2x \qquad (4\text{-}7.9)$$

(c) Particular Solution

The particular solution of equation (4-7.7) is obtained as follows

$$g(t) = \frac{1}{\left(D^2 + 4\right)} 3A\sin x$$
$$= 3A \frac{1}{\left(-1^2 + 4\right)} \sin x = A\sin x \qquad (4\text{-}7.10)$$

(g) General Solution

From equations (4-7.9) and (4-7.10), we get

$$g(t) = C.F. + P.S.$$
$$= A\cos 2x + B\sin 2x + \sin x \qquad (4\text{-}7.11)$$

(e) Final Solution

The final solution is obtained from (4-7.5) and (4-7.11), the final general solution is

$$y(x) = f(x)g(x) = \left[A\cos 2x + B\sin 2x + \sin x\right]e^{\frac{x^2}{2}} \qquad (4\text{-}7.12)$$

4-3. Solution of linear second-order o.d.e. by substitution by unknown arbitrary variable parameters

This is a **generalization** of the previous method, but with the modification of using two known solutions of the homogenous part of the o.d.e. and two arbitrary unknown functions or parameters.

Given the differential equation

$$\frac{d^2y}{dx^2} + P(x)\frac{dy}{dx} + Q(x)y = F(x) \qquad (4\text{-}8.1a)$$

Its homogeneous part is

$$\frac{d^2y}{dx^2} + P(x)\frac{dy}{dx} + Q(x)y = 0 \qquad (4\text{-}8.1b)$$

(a) Substitution by two known solutions of the homogenous equation

Assume that the complementary function of the above equation takes the form

$$y(x) = g(x)y_1(x) + f(x)y_2(x) \qquad (4\text{-}8.2)$$

Where, g and f are unknown parameters and y1 and y2 are solution of the homogenous equation (4-8.1b).

Differentiate $y(x)$ twice and substitute by y and its derivatives in the homogeneous equation provided that some restraints are imposed as follows.

$$y' = gy_1' + (g'y_1 + f'y_2) + fy_2' \qquad (4\text{-}8.3)$$

Conditioned on that the bracketed terms of free y_1 and y_2 vanish as follows

(b) Cancellation condition for simplification

$$\boxed{(g'y_1 + f'y_2) = 0 \qquad\qquad\qquad\qquad\qquad (4\text{-}8.4)}$$

Therefore, the second derivative becomes

$$y'' = gy_1'' + g'y_1' + f'y_2' + fy_2'' \qquad (4\text{-}8.5)$$

Substitute by y, y' and y" in the non-homogeneous equation (4-8.1a), we get

$$(gy_1'' + g'y_1' + f'y_2' + fy_2'') + P(x)(gy_1' + fy_2') + Q(x)y = F(x)$$

Or

$$g(y_1'' + Py_1' + Qy_1) + f(y_2'' + Py_2' + Qy_2) + g'y_1' + f'y_2' = F \qquad (4\text{-}8.6)$$

Where we have dropped all bracketed x arguments to simplify notation.

(c) Results of the two above assumptions

Since y1 and y2 are solutions of the homogenous equation (4-8.1b), therefore, the **bracketed terms in equation (4-8.6) vanish**. Thus,

$$g(0)+f(0)+g'y_1'+f'y_2'= F \qquad (4\text{-}8.7)$$

We therefore have two **simultaneous differential equations** (4-8.4) and (4-8.7), which provide the solution for the initial non-homogeneous equation

$$g'y_1 +f'y_2 = 0 \qquad (4\text{-}8.8a)$$
$$g'y_1'+f'y_2'= F \qquad (4\text{-}8.8b)$$

Since g' and f' are non-vanishing, therefore, the determinant of the above system of simultaneous differential equations is the Wronskian, not equal to zero. Integrating g' and f' gives g and f, as follows

$$g = \int g'dx + A$$
$$f = \int f'dx + A \qquad (4\text{-}8.9)$$

Substituting from (4-8.9) in (4-8.2), we get

$$y(x) = y_1\left(\int g'dx + A\right)+y_2\left(\int f'dx + B\right)$$
$$= Ay_1 + By_2 +\left(\int g'dx + \int f'dx\right) \qquad (4\text{-}8.10)$$
$$= \quad C.F. \quad + \quad P.S.$$

Equation (4-8.10) summarizes the intent of the present method, which is determining the **particular solution** of the non-homogenous second-order o.d.e. when two complementary functions y_1 and y_2 are known and when other methods become cumbersome.

4-4. Solution of linear nth-order o.d.e. by substitution by unknown arbitrary variable parameters

Farther **generalization** of the previous method applies to linear differential equations of orders higher than 2 and entails using n- known solutions of the homogenous parts of the o.d.e.'s and n-arbitrary unknown functions or parameters.

Given the differential equation

$$\sum_{i=0}^{n} P_i(x)\frac{d^i y}{dx^i} = F(x) \qquad (4\text{-}9.1a)$$

171

Its homogeneous part is

$$\sum_{i=0}^{n} P_i(x)\frac{d^i y}{dx^i} = 0 \qquad\qquad (4\text{-}9.1b)$$

(a) Substitution by n-known solutions of the homogenous equation

Assume that the complementary function of the above equation takes the form

$$y(x) = \sum_{i=1}^{n} g_i(x)y_i(x) \qquad\qquad (4\text{-}9.2)$$

Where, g_i are unknown parameters and y_i are solution of the homogenous equation (4-9.1b).

On differentiation, we get

$$y' = \sum_{i=1}^{n} g_i y_i' + \sum_{i=1}^{n} g_i' y_i \qquad\qquad (4\text{-}9.3)$$

Here, the imposed condition is that the term free y_i vanishes.

(b) First cancellation condition for simplification

$$\sum_{i=1}^{n} g_i' y_i = 0 \qquad\qquad (4\text{-}9.4)$$

Therefore, the second derivative becomes

$$y'' = \sum_{i=1}^{n} g_i y_i'' + \sum_{i=1}^{n} g_i' y_i' \qquad\qquad (4\text{-}9.5)$$

(c) Second cancellation condition for simplification

We repeat the same condition on the term of single derivatives such that

$$\sum_{i=1}^{n} g_i' y_i' = 0 \qquad\qquad (4\text{-}9.6)$$

We continue $(n-1)$ inclusive elimination process, which provides us **with n-simultaneous differential equations** solvable for n-parameters C's, and hence n-constants for the particular solution.

The repetitive conditioned vanishing sums is written as follows

$$\sum_{i=1}^{n} g_i' \frac{d^k y_i}{dx^k} = 0, \qquad (k = 0,1,2,...,n-2) \qquad (4\text{-}9.7)$$

Hence, we have (n-1) equations imposed on the arbitrary parameters in order to create (n-1) simultaneous o.d.e., in addition to the remaining equation of the highest order derivative, which makes n equations.

We sum up the vanishing equations and the expressions for derivatives of the general solution as follows:

Imposed condition on arbitrary parameters	Derivatives of general solution
	$y(x) = \sum_{i=1}^{n} g_i(x) y_i(x)$
$\sum_{i=1}^{n} g_i' y_i = 0$	$y' = \sum_{i=1}^{n} g_i y_i'$
$\sum_{i=1}^{n} g_i' y_i' = 0$	$y'' = \sum_{i=1}^{n} g_i y_i''$
$\sum_{i=1}^{n} g_i' \frac{d^k y_i}{dx^k} = 0, \qquad (k = 0,1,2,...,n-2)$	$y^{(k-1)} = \sum_{i=1}^{n} g_i \frac{d^{k-1} y_i}{dx^{k-1}} = 0, \quad (k = 1,2,...,n)$

Example 76

Solve the differential equation by the method of variable parameters

$$\frac{d^2 y}{dx^2} - \frac{1}{x}\frac{dy}{dx} = x \qquad (4\text{-}10.1)$$

Solution

The homogeneous part of the above equation is written as

$$y'' - \frac{1}{x} y' = 0 \qquad (4\text{-}10.2)$$

This can be solved by integration of each side with respect to each side's respective integrand, as follows

$$\frac{y''}{y'} = \frac{1}{x}$$

$$\frac{\frac{dy'}{dx}}{y'} = \frac{1}{x}$$

$$\frac{dy'}{y'} = \frac{dx}{x} \qquad (4\text{-}10.3a)$$

$$\ln y' = \ln x + \ln C$$

$$y' = Cx$$

Second integration gives

$$dy = Cdx$$

$$y = C_1 x^2 + C_2 \qquad (4\text{-}10.3b)$$

In order for y, equation (4-10.3b), to represent the solution of the non-homogenous equation (4-10.1), we apply the conditions given in equations (4-8.8) assuming that

$$y_1 = x^2$$
$$y_2 = 1$$
$$g = C_1 \qquad (4\text{-}10.4)$$
$$f = C_2$$

Therefore, equations (4-8.8), with substitution from equation (4-10.1), become

$$C_1'x^2 + C_2'.1 = 0$$
$$C_1'2x + C_2'0 = x \qquad (4\text{-}10.5)$$

C_1' and C_2' are derivatives of C_1 and C_2.

Solving the two simultaneous o.d.e.'s, we get

$$C_2' = -\frac{1}{2}x^2$$
$$C_1' = \frac{1}{2} \qquad (4\text{-}10.6)$$

Integrating, we get

$$C_2 = -\frac{1}{6}x^3 + A$$
$$C_1 = \frac{1}{2}x + B \qquad (4\text{-}10.7)$$

Therefore, the general solution, equation (4-10.3b), becomes

174

$$y = \left(\frac{1}{2}x + B\right)x^2 + \left(-\frac{1}{6}x^2 + A\right)C_2 \qquad (4\text{-}10.8a)$$

Or

$$y = \frac{1}{2}x^3 + Bx^2 + A \qquad (4\text{-}10.8b)$$

Example 77

Solve the differential equation by the method of variable parameters

$$\frac{d^2y}{dx^2} + y = \frac{1}{\cos x} \qquad (4\text{-}11.1)$$

Solution

The homogeneous part of the above equation is written as

$$(D^2 + 1)y = 0 \qquad (4\text{-}11.2)$$

Since the two roots of the characteristic equation are imaginary and distinct, the complementary function is

$$y = C_1 \cos x + C_2 \sin x \qquad (4\text{-}11.3)$$

In order for y, equation (4-11.3), to represent the solution of the non-homogenous equation (4-10.1), we apply the conditions given in equations (4-8.8) assuming that

$$
\begin{aligned}
y_1 &= \cos at \\
y_2 &= \sin at \\
g &= C_1 \\
f &= C_2
\end{aligned}
\qquad (4\text{-}11.4)
$$

Therefore, equations (4-8.8), with substitution from equation (4-11.1), become

$$
\begin{aligned}
C_1'\cos x + C_2'\sin x &= 0 \\
-C_1'\sin x + C_2'\cos x &= \frac{1}{\cos x}
\end{aligned}
\qquad (4\text{-}11.5)
$$

C_1' and C_2' are derivatives of C_1 and C_2.

Solving the two simultaneous o.d.e.'s, we get

$$C_1'\cos x \sin x + C_2'\sin^2 x = 0$$
$$\underline{-C_1'\sin x \cos x + C_2'\cos^2 x = 1}$$
$$C_2' = 1$$
$$C_1' = -\tan x$$

(4-11.6)

Integrating, we get

$$C_2 = x + A$$
$$C_1 = -\int \tan x dx = \ln \cos x + B$$

(4-11.7)

Therefore, the general solution, equation (4-11.3), becomes

$$y = (\ln \cos x + B)\cos x + (x + A)\sin x$$

(4-11.8a)

Or

$$y = A\sin x + B\cos x + \cos x \ln \cos x + x \sin x$$

(4-11.8b)

Example 78

Solve the differential equation by the method of variable parameters

$$\ddot{x} + a^2 x = f(t)$$

(4-12.1)

Solution

The homogeneous part of the above equation is written as

$$(D^2 + a^2)y = 0$$

(4-12.2)

Since the two roots of the characteristic equation are imaginary and distinct, the complementary function is

$$x = C_1 \cos at + C_2 \sin at$$

(4-12.3)

In order for y, equation (4-12.3), to represent the solution of the non-homogenous equation (4-10.1), we apply the conditions given in equations (4-8.8) assuming that

176

$$y_1 = \cos x$$
$$y_2 = \sin x$$
$$g = C_1$$
$$f = C_2$$

(4-12.4)

Therefore, equations (4-8.8), with substitution from equation (4-12.1), become

$$C_1{}'\cos at + C_2{}'\sin at = 0$$
$$- aC_1{}'\sin at + aC_2{}'\cos at = f(t)$$

(4-12.5)

C_1' and C_2' are derivatives of C_1 and C_2.

Solving the two simultaneous o.d.e.'s, we get

$$C_1{}'\cos at \sin at + C_2{}'\sin^2 at = 0$$

$$- C_1{}'\sin at \cos at + C_2{}'\cos^2 at = \frac{1}{a}f(t)\cos at$$
————————————————————

(4-12.6)

$$C_2{}' = \frac{1}{a}f(t)\cos at$$

$$C_1{}' = -\frac{1}{a}f(t)\sin at$$

Integrating, we get

$$C_2 = \frac{1}{a}\int \left(f(t)\cos at\right)dt + A$$

$$C_1 = -\frac{1}{a}\int \left(f(t)\sin at\right)dt + B$$

(4-12.7)

Therefore, the general solution, equation (4-12.3), becomes

$$y = \left(-\frac{1}{a}\int \left(f(t)\sin at\right)dt + B\right)\cos at + \left(\frac{1}{a}\int \left(f(t)\cos at\right)dt + A\right)\sin at$$

(4-12.8a)

Or

$$y = A\sin at + B\cos at - \cos at\int \left(f(t)\sin at\right)dt + \sin at\int \left(f(t)\cos at\right)dt$$

(4-12.8b)

The integration terms can be arranged by changing the differentials and arranging, as follows

$$y = A\sin at + B\cos at + \frac{1}{a}\left[\cos at\int_0^t \big(f(t)\sin au\big)du - \sin at\int_0^t \big(f(t)\cos au\big)du\right]$$

$$= A\sin at + B\cos at + \frac{1}{a}\left[\int_0^t f(t)\big(\cos at\sin au - \sin at\cos au\big)du\right] \tag{4-12.9}$$

$$= A\sin at + B\cos at + \frac{1}{a}\int_0^t f(t)\sin a(t-u)du$$

The final solution shows the **superposition** of the complementary solution, which is a sinusoidal wave solution of the homogenous **wave equation**, and the particular solution, which is the source term of integral that distorts the sinusoidal wave in proportion to the time-phase shift (t-u) and the magnitude of f(t).

4-5. Solution of linear second-order o.d.e. by changing of independent variable

Another modification of the previous methods emphasizes reducing **variables coefficients of unknown dependent variable**, y, by changing the known **independent variable**, x, by conditional substitution as follows.

Given the differential equation

$$\frac{d^2y}{dx^2} + P(x)\frac{dy}{dx} + Q(x)y = F(x) \tag{4-13.1}$$

We will change the independent variable x with unknown function that will be determined so as to make the variable coefficients of y and its derivatives constants.

(a) Substitution of independent variable

$$t = g(x) \tag{4-13.2}$$

Next is to express the derivatives of y in terms of t as follows

$$\frac{dy}{dx} = \frac{dy}{dt}\frac{dt}{dx} = \frac{dy}{dt}\frac{dg}{dx}$$
$$\frac{d^2y}{dx^2} = \frac{d^2y}{dt^2}\left(\frac{dg}{dx}\right)^2 + \frac{dy}{dt}\frac{d^2g}{dx^2} \tag{4-13.3}$$

Substituting by the derivatives of y in (4-13.1), we get

178

$$\frac{d^2y}{dt^2}\left(\frac{dg}{dx}\right)^2 + \frac{dy}{dt}\frac{d^2g}{dx^2} + P\frac{dy}{dt}\frac{dg}{dx} + Qy = F \qquad (4\text{-}13.4)$$

This can be written in a form that facilitates our choice of g(x) or its derivatives, as follows

$$\frac{d^2y}{dt^2} + \frac{dy}{dt}\frac{\left(\dfrac{d^2g}{dx^2} + P\dfrac{dg}{dx}\right)}{\left(\dfrac{dg}{dx}\right)^2} + \frac{Q}{\left(\dfrac{dg}{dx}\right)^2}y = \frac{F}{\left(\dfrac{dg}{dx}\right)^2} \qquad (4\text{-}13.5)$$

(b) Determining the unknown substitution function

1. Fixing the coefficient of y

Equating the coefficient of y, in equation (4-13.5) by constant, as follows

$$\frac{Q}{\left(\dfrac{dg}{dx}\right)^2} = \frac{1}{C^2} \qquad (4\text{-}13.6a)$$

Therefore,

$$\frac{dg}{dx} = C\sqrt{Q} \qquad (4\text{-}13.6b)$$

Integrating, we get

$$g = \int C\sqrt{Q}\,dx \qquad (4\text{-}13.7)$$

2. Examining the coefficient of y' for fixation

We do not know whether fixing the coefficient of y would fix the coefficient of its derivative y'. Therefore, the obtained g and its derivatives, equation (4-13.7) is substituted in the coefficient of y', equation (4-13.5).

First, differentiate g, as follows

$$g = \int C\sqrt{Q}\,dx$$
$$\frac{dg}{dx} = C\sqrt{Q} \qquad (4\text{-}13.8)$$
$$\frac{d^2g}{dx^2} = C\frac{1}{2\sqrt{Q}}\frac{dQ}{dx}$$

Therefore, the coefficient of y', in equation (4-13.5), becomes

$$\frac{\left(\dfrac{d^2g}{dx^2}+P\dfrac{dg}{dx}\right)}{\left(\dfrac{dg}{dx}\right)^2}=\frac{\left(C\dfrac{1}{2\sqrt{Q}}\dfrac{dQ}{dx}+PC\sqrt{Q}\right)}{QC^2}$$

$$=\frac{\left(\dfrac{1}{2}\dfrac{dQ}{dx}+PQ\right)}{CQ\sqrt{Q}}$$

(4-13.9)

This provides the condition for reducing the variables coefficients of y and y' onto constants as follows

$$\frac{\left(\dfrac{1}{2}\dfrac{dQ}{dx}+PQ\right)}{Q\sqrt{Q}}=A$$

(4-13.10)

If there exist a constant A, defined the above equation, then g(x) could be determined by substituting that constant in equation (4-13.7), which gives the new independent variable, t.

Example 79

Use the method of change of the independent variable x to solve the o.d.e.

$$x\frac{d^2y}{dx^2}-2(x^3+1)\frac{dy}{dx}+x^5=x^5e^{x^3}$$

(4-14.1)

Solution

We will change the independent variable x with unknown function that will be determined to make the variable coefficients of y and its derivatives constants.

Equation (4-14.1) is divided by x and gives the previous defined P(x), Q(x), and F(x), as follows

$$P(x)=-2\left(\frac{x^3+1}{x}\right)$$

$$Q(x)=x^4$$

$$F(x)=x^4e^{x^3}$$

(4-14.2)

Equation (4-13.7), with the substitution by Q from the above equation, gives the transformation to the new independent variable t, as follows

$$g = \int C\sqrt{Q}\,dx$$
$$= C\int \sqrt{x^4}\,dx \qquad\qquad (4\text{-}14.3)$$
$$= \frac{C}{3}x^3$$

Let us see whether equation (4-13.10) can be satisfied. Substitute from (4-14.2) into (4-13.10), we get

$$\frac{\left(\dfrac{1}{2}\dfrac{dQ}{dx} + PQ\right)}{CQ\sqrt{Q}} = \frac{\left(2x^3 - 2\dfrac{x^3+1}{x}x^4\right)}{Cx^6}$$
$$= \frac{\left(-2x^6\right)}{Cx^6} = -\frac{2}{C} \qquad\qquad (4\text{-}14.4)$$

Thus, if we set $C = 3$, then coefficient of y', equation (4-13.5) takes the form

$$\frac{d^2y}{dt^2} - \frac{2}{3}\frac{dy}{dt} + \frac{1}{9}y = \frac{1}{9}e^t \qquad\qquad (4\text{-}14.5)$$

This is the transformed equation of constant coefficients equivalent to equation (4-14.1) with the new variable t defined by g, equation (4-14.3).

The characteristic equation of the above equation is

$$9m^2 - 6m + 1 = 0 \qquad\qquad (4\text{-}14.6)$$

Hence, we have two real equal roots $m_1 = m_2 = 1/3$ leading the **complementary function**

$$y = (A + Bt)e^{\frac{1}{3}t} \qquad\qquad (4\text{-}14.7)$$

The **particular solution** of (4-13.5) is

$$y = \frac{1}{(3D-1)^2}e^t$$

$$= \frac{1}{(3-1)^2}e^t \qquad (4\text{-}14.8)$$

$$= \frac{1}{4}e^t$$

The **general solution** is

$$G.S. = C.F. + P.S.$$

$$y = (A + Bt)e^{\frac{1}{3}t} + \frac{1}{4}e^t \qquad (4\text{-}14.9)$$

4-6. Relation between the solutions of a second-order homogenous linear o.d.e.

The **linear independence** the two particular solutions of homogeneous linear o.d.e. of the second order allows the determination of one of the two particular solutions if the other is known.

For, if y_1 and y_2 are two particular solutions, given y_1, the other solution, y_2, is determined as follows.

Given the linear homogenous second-order o.d.e.

$$\frac{d^2y}{dx^2} + P(x)\frac{dy}{dx} + Q(x)y = 0 \qquad (4\text{-}15.1)$$

The linear independence of y_1 from y_2, implies non-vanishing **Wronskian**

$$W[y_1, y_2] = \begin{vmatrix} y_1 & y_2 \\ \dfrac{dy_1}{dx} & \dfrac{dy_2}{dx} \end{vmatrix} = y_1\frac{dy_2}{dx} - y_2\frac{dy_1}{dx} \neq 0 \qquad (4\text{-}15.2a)$$

The derivative of the Wronskian will be needed due to its similar structure to the above, as follows

$$\frac{dW[y_1, y_2]}{dx} = y_1\frac{d^2y_2}{dx^2} - y_2\frac{d^2y_1}{dx^2} \qquad (4\text{-}15.2b)$$

Thus, the general solution is

$$y = C_1 y_1 + C_2 y_2 \qquad (4\text{-}15.3)$$

Substituting by each solution in the homogenous equation (4-15.1), we get

$$\frac{d^2 y_1}{dx^2} + P(x)\frac{dy_1}{dx} + Q(x)y_1 = 0$$

$$\frac{d^2 y_2}{dx^2} + P(x)\frac{dy_2}{dx} + Q(x)y_2 = 0 \qquad (4\text{-}15.4)$$

Eliminate Q between the two equations, we get

$$\left(y_2 \frac{d^2 y_1}{dx^2} - y_1 \frac{d^2 y_2}{dx^2} \right) + P(x)\left(y_2 \frac{dy_1}{dx} - y_1 \frac{dy_2}{dx} \right) = 0 \qquad (4\text{-}15.5)$$

Substituting by W and its derivative from (4-15.2), we get

$$\frac{dW[y_1, y_2]}{dx} + P(x)W[y_1, y_2] = 0 \qquad (4\text{-}15.6)$$

Separating and integrating, we get

$$\frac{dW[y_1, y_2]}{W[y_1, y_2]} = -P(x)dx$$

$$\ln(W[y_1, y_2]) = -\int P(x)dx + \ln C \qquad (4\text{-}15.7)$$

Thus,

$$W[y_1, y_2] = Ce^{-\int P(x)dx} \qquad (4\text{-}15.8)$$

Substituting by W in (4-15.2a), we get

$$y_1 \frac{dy_2}{dx} - y_2 \frac{dy_1}{dx} = Ce^{-\int P(x)dx} \qquad (4\text{-}15.9)$$

Equation (4-15.9) enables us determine y_2 in terms of y_1, as follows.

Divide (4-15.9) by y_1^2, we get

$$\frac{y_1 \dfrac{dy_2}{dx} - y_2 \dfrac{dy_1}{dx}}{y_1^{\,2}} = \frac{Ce^{-\int P(x)dx}}{y_1^{\,2}} \qquad (4\text{-}15.10)$$

This can be written as follows

$$\frac{d}{dx}\left(\frac{y_2}{y_1}\right) = \frac{Ce^{-\int P(x)dx}}{y_1^{\,2}} \qquad (4\text{-}15.11)$$

Which upon integration, gives

$$\frac{y_2}{y_1} = \int \frac{Ce^{-\int P(x)dx}}{y_1^{\,2}}dx + A \qquad (4\text{-}15.12)$$

Since we are seeking particular solution, we could set the arbitrary constants as C=1 and A =0. Also, since the above quotient is not a constant, therefore, y_1 and y_2 are linearly independent.

Therefore, we obtain the unknown particular solution y_2 in terms of y_1 as follows

$$y_2 = y_1 \int \frac{e^{-\int P(x)dx}}{y_1^{\,2}}dx \qquad (4\text{-}15.13)$$

Therefore, the general solution for equation (4-15.1) is a linear combination of y_1 and y_2 given by

$$y = y_1\left(C_1 + C_2\int \frac{e^{-\int P(x)dx}}{y_1^{\,2}}dx\right) \qquad (4\text{-}15.14)$$

Example 80

Find the general solution of the equation

$$(1-x^2)\frac{d^2y}{dx^2} - 2x\frac{dy}{dx} + 2y = 0 \qquad (4\text{-}16.1)$$

Solution

A particular solution for the above equation can be deduced visually as $y_1 = x$. The second particular solution y_2, is obtained by equation (4-15.13) as follows.

The coefficient of y' is obtained by dividing the above equation by $(1-x^2)$

184

$$P(x) = -\frac{2x}{(1-x^2)} \qquad (4\text{-}16.2)$$

And

$$\int P(x) = -\frac{2x}{(1-x^2)} dx$$

$$= \frac{d(1-x^2)}{(1-x^2)} = \ln(1-x^2) \qquad (4\text{-}16.3)$$

Therefore, equation (4-15.13) becomes

$$y_2 = y_1 \int \frac{e^{-\ln(1-x^2)}}{y_1^2} dx$$

$$= x \int \frac{e^{-n(1-x^2)}}{x^2} dx \qquad (4\text{-}16.4)$$

$$= x \int \frac{dx}{x^2(1-x^2)}$$

Writing the integrand in its partial fractions, we get

$$\frac{1}{x^2(1-x^2)} = \frac{A}{x^2} + \frac{B}{(1-x)} + \frac{C}{(1+x)}$$

$$A(1-x^2) + Bx^2(1+x) + Cx^2(1-x) = 1 \qquad (4\text{-}16.5)$$

$$A = 1$$

$$B = C = \frac{1}{2}$$

$$y_2 = x \int \left(\frac{dx}{x^2} + \frac{dx}{2(1-x)} + \frac{dx}{2(1+x)} \right)$$

$$= x \int \left(-\frac{1}{x} - \frac{1}{2}\ln(1-x) + \frac{1}{2}\ln(1+x) \right) \qquad (4\text{-}16.6)$$

$$= -1 + \frac{x}{2}\ln\frac{1+x}{1-x}$$

The general solution is

$$y = C_1 x + C_2 \left(-1 + \frac{x}{2}\ln\frac{1+x}{1-x} \right) \qquad (4\text{-}16.7)$$

185

4-7. Exercises on second-order linear o.d.e. with variable coefficients

Solve the following o.d.e.

4-1.   Examine the linearity of the following relationships:

a) $x,$ $x+1$

b) $x^2,$ $-2x^2$

c) $0,$ 1 x

d) $x,$ $x+1$ $x+2$

e) $x,$ $x^2,$ x^3

f) $e^x,$ $e^{2x},$ e^{3x}

g) $\sin x,$ $\cos x,$ 1

g) $\sin^2 x,$ $\cos^2 x,$ 1

--- --- --- ---

[Ans: a) no   b) yes   c) yes   d) yes   e) no   f) no   g) no   h) yes]

4-2.   Solve the o.d.e. by reduction of order given a particular solution P.S. = sinx / x

$$\frac{d^2y}{dx^2}+\frac{1}{x}\frac{dy}{dx}+y=0 \quad [\text{Ans}: \ y=\frac{1}{x}(A\cos x+B\sin x)]$$

4-3.   Solve the o.d.e. by reduction of order given a particular solution P.S. = x

$$x^2(\ln x-a)\frac{d^2y}{dx^2}-x\frac{dy}{dx}+y=0 \qquad [\text{Ans}: \ y=Ax+B\ln x]$$

4-4.   Show that y = sinx /x is one of the solutions of the complementary function of the o.d.e. given below, and hence find its general solution

$$\frac{d^2y}{dx^2} + \frac{2}{x}\frac{dy}{dx} + y = \frac{2}{x}\cos x \qquad [\text{Ans}: \quad y = \frac{1}{x}(A\cos x + B\sin x) + \sin x]$$

4-5.   Show that $y = 1/(1-x^2)$ is one of the solutions of the complementary function of the o.d.e. given below, and hence find its general solution

$$(1-x)^2\frac{d^2y}{dx^2} - 4x\frac{dy}{dx} - 2y = 1 \qquad [\text{Ans}: \quad y = \frac{A+Bx}{1-x^2} = \frac{1}{2}]$$

4-6.   Show that $y = \cos x$ is a solution of the o.d.e

$$y'' + 3y'\tan x + (1 + 3\tan^2 x)y = 0$$

Hence, find the complete integral of the equation

$$y'' + 3y'\tan x + (1 + 3\tan^2 x)y = \cos^2 x$$

$$\left[\text{Ans}: \quad x = A\sin x + b\cos x + x\sin x\cos x + \cos^2 x \right]$$

4-7.   By reduction to **normal form**, solve the o.d.e.

$$\frac{d^2y}{dx^2} + \frac{1}{x}\frac{dy}{dx} - \frac{y}{4x^2} = 0 \qquad \left[\text{Ans}: \quad y = \frac{Ax+B}{\sqrt{x}} \right]$$

4-8.   By reduction to **normal form**, solve the o.d.e.

$$\frac{d^2y}{dx^2} + 8x^3\frac{dy}{dx} + (16x^6 + 12x^3 + 1)y = xe^{-x^4}$$

$$\left[\text{Ans}: \quad x = e^{-x^4}(A\cos x + B\sin x + x) \right]$$

4-9.   By reduction to **normal form**, solve the o.d.e.

$$x^2\frac{d^2y}{dx^2} - 4x\frac{dy}{dx} + (6 + x^2)y = x^7 \qquad \left[\text{Ans}: \quad y = x^2(A\cos x + B\sin x + x^3 - 6x) \right]$$

4-10.   By reduction to **normal form**, solve the o.d.e. for finite solution at x=0c

$$x\frac{d^2y}{dx^2} + 2(1+x^2)\frac{dy}{dx} + (4x+x^3)y = 0 \qquad \left[\text{Ans}: \ y = e^{-\frac{1}{2}x^2}\frac{\sin x}{x}\right]$$

Solve the following non-homogenous o.d.e. equations by variation of parameters

4-11.
$$x^2\frac{d^2y}{dx^2} - x\frac{dy}{dx} = 3x^3 \qquad \left[\text{Ans}: \ y = A + Bx^2 + x^3\right]$$

4-12.
$$x^2\frac{d^2y}{dx^2} + x\frac{dy}{dx} - y = x^2 \qquad \left[\text{Ans}: \ y = \frac{1}{3}x^2 + Ax + \frac{B}{x}\right]$$

4-13.
$$\frac{d^3y}{dx^3} + \frac{dy}{dx} = \sec x$$
$$\left[\text{Ans}: \ y = A + Bx + Cx^2 + e^x(E + Fx + Gx^2) - \frac{1}{2}x^3 - \frac{1}{24}x^4\right]$$

4-14.
$$\frac{d^2y}{dx^2} + y = \sec 2x\sqrt{\sec 2x} \qquad \left[\text{Ans}: \ y = A\cos x + B\sin x - \sqrt{\cos 2x}\right]$$

4-15.
$$\frac{d^2y}{dx^2} + y = \csc^2 x \qquad \left[\text{Ans}: \ y = A\cos x + B\sin x + \frac{1}{2}\cos x\cot x\right]$$

4-16. Solve the o.d.e. $\quad \dfrac{d^2y}{dx^2} + y = 1 - \csc x$
$$\left[\text{Ans}: \ y = A\cos x + B\sin x + 1 + x\cos x - \sin x \ln \sin x\right]$$

4-17. Find the complete integral of the equation
$$\frac{d^2y}{dx^2} - 2y = 4x^2 e^{x^2} \qquad \left[\text{Ans}: \ y = Ae^{\sqrt{2}x} + Be^{-\sqrt{2}x} + e^{x^2}\right]$$

4-18. $\quad \dfrac{d^2y}{dx^2} + (\tan x + 2\cos x)\dfrac{dy}{dx} + y\cos^2 x = 2\cos^2 x$

$$\left[\ \text{Ans}:\quad y = (A + B\sin x)e^{-\sin x} + 2\ \right]$$

Find a suitable substitution of independent variable x to solve the equations

4-19.

$$x^5\dfrac{d^2y}{dx^2} + 2x^4\dfrac{dy}{dx} + 4xy = 2 \qquad\qquad \left[\ \text{Ans}:\quad y = A\cos\dfrac{1}{x} + B\sin\dfrac{2}{x} + \dfrac{1}{2x}\ \right]$$

4-20. $\quad \sin^2 x\dfrac{d^2y}{dx^2} + \sin\cos x\dfrac{dy}{dx} - 4y = 0$

$$\left[\ \text{Ans}:\quad y = A(\operatorname{cosec}x - \cot x)^2 + \dfrac{B}{(\operatorname{cosec}x - \cot x)^2}\ \right]$$

4-21. $\quad \dfrac{d^2y}{dx^2} + \tan x\dfrac{dy}{dx} + 4y\cos^2 x = 8\sin x\cos^2 x \quad$, when $x = 0$ and $y' = 0$

$$\left[\ \text{Ans}:\quad y = 2\sin x - \sin(2\sin x)\ \right]$$

4-22. $\qquad$ Given a particular solution $y_1 = \sin x\,/\,x$ of the equation below, find the other particular solution, y_2.

$$x\dfrac{d^2y}{dx^2} + 2\dfrac{dy}{dx} + xy = 0 \qquad\qquad \left[\ \text{Ans}:\quad y_2 = \dfrac{\cos x}{x}\ \right]$$

4-23. $\qquad$ Given a particular solution $y_1 = x$ of the equation below, find the other particular solution, y_2.

$$x^2(\ln x - 1)\dfrac{d^2y}{dx^2} - x\dfrac{dy}{dx} + y = 0 \qquad\qquad \left[\ \text{Ans}:\quad y_2 = \ln x\ \right]$$

4-24. $\qquad$ Given a particular solution $y_1 = \sqrt{x}$ of the equation below, find the other particular solution, y_2.

$$4x^2 \frac{d^2y}{dx^2} + 4x \frac{dy}{dx} - y = 0 \qquad\qquad \left[\text{Ans}: \ y_2 = \frac{1}{\sqrt{x}} \right]$$

4-25.　　　Given a particular solution $y_1 = \cos x$ of the equation below, find the other particular solution, y_2.

$$\frac{d^2y}{dx^2} + 3 \tan x \frac{dy}{dx} + (1 + 3 \tan^2 x)y = 0 \qquad \left[\text{Ans}: \ y_2 = \sin 2x \right]$$

4-26.　　　Given a particular solution $y_1 = \sin x$ of the equation below, find the other particular solution, y_2.

$$x^2 \frac{d^2y}{dx^2} - 2x \frac{dy}{dx} + (x^2 + 2)y = 0 \qquad \left[\text{Ans}: \ y_2 = x \cos x \right]$$

4-27.　　　Given a particular solution $y_1 = \tan x$ of the equation below, find the other particular solution, y_2.

CHAPTER 5

SOLUTION OF O.D.E.
BY SERIES EXPANSIONS

5-1. Existence of series expansion solution

(a) Linear combinations of linearly independent particular solutions of homogeneous o.d.e.

We have shown, in equation (3-12.2), that an ordinary differential equation of the nth order of the form

$$\sum_{n=0}^{n} a_n(x) \frac{d^n y(x)}{dx^n} = f(x) \tag{5-1.1}$$

can be solved by the solution of n **simultaneous** linear differential equations as follows

$$D^j \left(\sum_{i=1}^{n} C_i y_i(x) \right) = 0 \tag{5.1.2}$$

which **Wronskian determinant**, at every point x, in the interval $a \le x \le b$, vanishes as follows

$$W(y_i(x)) = \begin{vmatrix} y_i & y_2 & .. & y_n \\ y_i' & y_2' & .. & y_n' \\ .. & .. & .. & .. \\ y_i^{n-1} & y_2^{n-1} & .. & y_n^{n-1} \end{vmatrix} = 0 \tag{5.1.3}$$

The n particular solutions y_i are **linearly independent** combined by the constants C_i.

$$y = \sum_{i=1}^{n} C_i y_i(x) \qquad\qquad (5.1.4)$$

(b) Analyticity of solution

Given the second-order linear o.d.e. (linear in y and its derivatives, not in x)

$$a_0(x)\frac{d^2y(x)}{dx^2} + a_1(x)\frac{dy(x)}{dx} + a_2(x)y(x) = 0 \qquad\qquad (5-2.1)$$

Also, given that

$a_0(x)$, $a_1(x)$, and $a_2(x)$ are **analytical function**

(determined by convergent solutions at given point) such that at

$x = \varepsilon$ and $a_0(\varepsilon) \neq 0$,

Then

There exists a series expansion solution in the neighborhood of $x = \varepsilon$, given by

$$y = c_0 + c_0(x - \varepsilon) + c_1(x - \varepsilon)^2 + \dots + c_n(x - \varepsilon)^n + \dots \qquad\qquad (5-2.2)$$

The length of the expansion series depends on the desired accuracy of the sought particular solution.

(c) Expansibility of solution

Given the second-order linear o.d.e. (5-2.1) and that

$a_0(x)$, $a_1(x)$, and $a_2(x)$ are **analytical function**

(determined by convergent solutions at given point) such that at

$x = \varepsilon$

Such that

For finite order   s   of $a_0(x)$ implies $a_0(\varepsilon) = 0$

192

For finite order (s-1) of $a_1(x)$ implies $a_1(\varepsilon) = 0$

For finite order (s-2) of $a_2(x)$ implies $a_2(\varepsilon) = 0$

Then

There exists at **least one non-trivial series expansion solution** in the neighborhood of $x = \varepsilon$, given by

$$y = c_0(x - \varepsilon)^s + c_1(x - \varepsilon)^{s+1} + \ldots\ldots + c_n(x - \varepsilon)^{s+n} + \ldots \tag{5-3}$$

Where s is some **real number**.

5-2. Solution by Taylor's series expansion

Given the function $y(x)$ and its derivatives, we could construct a solution of the form (5-2.1) at an arbitrary point $x = \varepsilon$ in the interval, where y and its derivatives are analytic.

Thus, function $y(x)$ is expanded as sum of powers

$$y = \sum_{n=0}^{\infty} C_n(x - \varepsilon)^n \tag{5.4.1}$$

The coefficients C_n's are determined by differentiating the above expansion series n times to get

$$y^{(0)} = \sum_{n=0}^{\infty} C_n(x - \varepsilon)^n$$

$$y^{(1)} = \sum_{n=1}^{\infty} nC_n(x - \varepsilon)^{n-1}$$

$$y^{(2)} = \sum_{n=2}^{\infty} n(n-1)C_n(x - \varepsilon)^{n-2} \tag{5.4.2a}$$

$$\ldots\ldots\ldots \quad \ldots\ldots\ldots \quad \ldots\ldots\ldots \quad .$$

$$y^{(k)} = \sum_{n=k}^{\infty} (n-k+)!C_n(x - \varepsilon)^{n-k}$$

We will aggregate the terms that contain independent variable x from the constants, and noting the factorial notations: $(0!) = 1$, $(1!)=1$, $(2!) =2(1)$, $(3!) =3(2)(1)$,....,as follows

$$y^{(0)} = (0!)C_0 + \sum_{n=1}^{\infty} C_n (x - \varepsilon)^n$$

$$y^{(1)} = (1!)C_1 + \sum_{n=2}^{\infty} nC_n (x - \varepsilon)^{n-1}$$

$$y^{(2)} = (2!)C_2 + \sum_{n=3}^{\infty} n(n-1)C_n (x - \varepsilon)^{n-2} \qquad (5.4.2b)$$

....

$$y^{(k)} = (k!)C_k + \sum_{n=k+1}^{\infty} [(n - k + 1)!]C_n (x - \varepsilon)^{n-k}$$

Putting $x = \varepsilon$ in these expressions gives

$$C_0 = \frac{y^{(0)}}{0!}$$

$$C_1 = \frac{y^{(1)}}{1!}$$

$$C_2 = \frac{y^{(2)}}{2!} \qquad (5.4.3)$$

$$C_k = \frac{y^{(k)}}{k!}$$

Substituting by the above constants in (5-4.1), we obtain Taylor's series expansion as follows:

$$y(x) = y(\varepsilon) + (x - \varepsilon)\frac{dy(x)}{dx}\Big|_{x=\varepsilon} + \frac{1}{2!}(x - \varepsilon)^2 \frac{d^2y(x)}{dx^2}\Big|_{x=\varepsilon} + .. + \frac{1}{n!}(x - \varepsilon)^n \frac{d^n y(x)}{dx^n}\Big|_{x=\varepsilon} + ... \qquad (5\text{-}4.4)$$

We have replaced the orders of differentiation by their explicit forms for clarity.

The abbreviated form is

$$y(x) = y(\varepsilon) + (x - \varepsilon)y^{(1)}(\varepsilon) + \frac{1}{2!}(x - \varepsilon)^2 y^{(2)}(\varepsilon) + .. + \frac{1}{n!}(x - \varepsilon)^n y^{(n)}(\varepsilon) + ... \qquad (5\text{-}4.5)$$

Clearly, the abbreviated form lacks the advantage of showing that the derivatives of x were first obtained by differentiation and then determined at the point $x = \varepsilon$.

Example 81

Find a power series solution of the equation

194

$$\frac{d^2y(x)}{dx^2} + 3x\frac{dy(x)}{dx} - 6y = 0 \qquad\qquad (5\text{-}5.1)$$

Subject to

$$(x, y, y') = (0, 1, 0.1) \qquad\qquad (5\text{-}5.2)$$

Solution

(a) First, let us construct few **higher-orders of derivatives** of y by differentiating equation (5-5.1) as follows

1st differentiation gives:

$$\frac{d^3y(x)}{dx^3} + 3\left(\frac{dy(x)}{dx} + x\frac{d^2y(x)}{dx^2}\right) - 6\frac{dy(x)}{dx} = 0 \qquad\qquad (5\text{-}5.3a)$$

2nd differentiation gives:

$$\frac{d^4y(x)}{dx^4} + 3\left(2\frac{d^2y(x)}{dx^2} + x\frac{d^3y(x)}{dx^3}\right) - 6\frac{d^2y(x)}{dx^2} = 0 \qquad\qquad (5\text{-}5.3b)$$

3rd differentiation gives:

$$\frac{d^5y(x)}{dx^5} + 3\left(3\frac{d^3y(x)}{dx^3} + x\frac{d^4y(x)}{dx^4}\right) - 6\frac{d^3y(x)}{dx^3} = 0 \qquad\qquad (5\text{-}5.3c)$$

nth differentiation gives:

$$\frac{d^n y(x)}{dx^n} + 3\left((n-2)\frac{d^{(n-2)}y(x)}{dx^{(n-2)}} + x\frac{d^{(n-1)}y(x)}{dx^{(n-1)}}\right) - 6\frac{d^{(n-2)}y(x)}{dx^{(n-2)}} = 0 \qquad\qquad (5\text{-}5.3d)$$

(b) Constructing the coefficients of Taylor's series

Starting by the initial conditions given in equation (5-5.2), we calculate the second derivative of y, from equation (5-5.1) by direct substitution as follows:

$$\left.\frac{d^2y(x)}{dx^2} + 3x\frac{dy(x)}{dx} - 6y\right|_{x=0} = \frac{d^2y(0)}{dx^2} + 3(0)(0.1) - 6(1) = 0$$

$$\frac{d^2y(0)}{dx^2} = 6 \qquad\qquad (5\text{-}5.4a)$$

Substituting by the above obtained value and the initial condition from (5-5.2) in (5-5.3a), we get

$$\left|\frac{d^3y(x)}{dx^3} + 3\left(\frac{dy(x)}{dx} + x\frac{d^2y(x)}{dx^2}\right) - 6\frac{dy(x)}{dx}\right|_{x=0} = \frac{d^3y(0)}{dx^3} + 3(0.1 + 0(6)) - 6(0.1) = 0$$

(5-5.4b)

$$\frac{d^3y(0)}{dx^3} = 0.3$$

Repeating the above substitution in (5-5.3b), we get

$$\left|\frac{d^4y(x)}{dx^4} + 3\left(2\frac{d^2y(x)}{dx^2} + x\frac{d^3y(x)}{dx^3}\right) - 6\frac{d^2y(x)}{dx^2}\right|_{x=0} = \frac{d^4y(0)}{dx^4} + 3(2(6) + 0(6)) - 6(6) = 0$$

(5-5.4c)

$$\frac{d^4y(0)}{dx^4} = 0$$

Repeating the above substitution in (5-5.3c), we get

$$\left|\frac{d^5y(x)}{dx^5} + 3\left(3\frac{d^3y(x)}{dx^3} + x\frac{d^4y(x)}{dx^4}\right) - 6\frac{d^3y(x)}{dx^3}\right|_{x=0} = \frac{d^5y(0)}{dx^5} + 3(3(0.3) + 0(0)) - 6(0.3) = 0$$

(5-5.4d)

$$\frac{d^5y(0)}{dx^5} = -0.9$$

(c) Constructed Taylor's series solution

Substituting

$$y(x) = y(\varepsilon) + (x - \varepsilon)\frac{dy(x)}{dx}\bigg|_{x=\varepsilon} + \frac{1}{2!}(x-\varepsilon)^2\frac{d^2y(x)}{dx^2}\bigg|_{x=\varepsilon} + .. + \frac{1}{n!}(x-\varepsilon)^n\frac{d^ny(x)}{dx^n}\bigg|_{x=\varepsilon} + ...$$

(5-5.5)

$$= y(0) + 0.1x + \frac{6}{2!}x^2 + \frac{0.3}{3!}x^3 - \frac{0.9}{5!}x^5 + \frac{8.1}{7!}x^7 + ..$$

(d) Level of correctness

Suppose that the obtained series expansion in (5-5.5) is required to represent the required solution y to less than 0.0001 up to the 5th term. Then the 6th terms should be set to

$$\frac{8.1}{7!}x_1^7 \le (0.0001)/2$$

(5-5.6)

Thus,

$$x_1^7 \le \frac{7!}{8.1}(0.0001)/2 = 0.031$$

(5-5.7)

$$x_1 \le 0.609119$$

Example 82

Find a power series solution of the equation

$$x\frac{d^2y(x)}{dx^2} + \frac{dy(x)}{dx} + xy = 0 \qquad (5\text{-}6.1)$$

In the neighborhood $x = 0$

Solution

(a) First, let us construct **recurrence formula** for the coefficients of Taylor's series by differentiating (5-6.1) n times

Each term in equation (5-6.1) can be differentiated n times to give

$$\frac{d^n}{dx^n}\left(xy''\right) = xy^{(n+2)} + ny^{(n+1)}$$
$$\frac{d^n}{dx^n}\left(y'\right) = y^{(n+1)} \qquad (5\text{-}6.2)$$
$$\frac{d^n}{dx^n}(xy) = xy^{(n)} + y^{(n-1)}$$

Where exponents denote order of derivatives.

Adding the above three and equating to zero, we get the (n+2)-order of equation (5-6.2)

$$xy^{(n+2)} + ny^{(n+1)} + y^{(n+1)} + xy^{(n)} + ny^{(n-1)} = 0 \qquad (5\text{-}6.3)$$

The recurrence formula for the given boundary condition, $x = 0$, is therefore,

$$ny^{(n+1)} + y^{(n+1)} + ny^{(n-1)} = 0 \qquad (5\text{-}6.4)$$

Equation (5-6.4) gives the coefficients in Taylor's series

197

$$y^{(n+1)} = -\frac{n}{n+1}y^{(n-1)}, \qquad n = 1, 2, \ldots \infty \qquad\qquad (5\text{-}6.5)$$

We note that we have **two unknown constants,** $y^{(0)}$ and $y^{(1)}$, required to determine the final series expansion.

First, we need the initial value y at x = 0, which corresponds to $y^{(0)}$ in the above recurrence formula.

Second, we initial value of y' as x = 0, since the recurrence formula in (5-6.5) skips one order of differentiation, jumps from (n-1) to (n+2).

Determination of the coefficients of Taylor's series

Starting from the two recurrence formulas

$$y^{(n+1)} = -\frac{n}{n+1}y^{(n-1)}, \qquad n = 1, 2, \ldots \infty \qquad\qquad (5\text{-}6.6)$$

$$y^{(n+2)} = -\frac{n+1}{n+2}y^{(n)}, \qquad n = 0, 1, 2, \ldots \infty \qquad\qquad (5\text{-}6.7)$$

Varying n from 1 to ∞, skipping even numbers, we get

$$y^{(n+1)} = -\frac{n}{n+1}y^{(n-1)}$$

$$y^{(2)} = -\frac{1}{2}y^{(0)}$$

$$y^{(4)} = -\frac{3}{4}y^{(2)} = \frac{3}{8}y^{(0)} \qquad\qquad (5\text{-}6.8)$$

$$y^{(6)} = -\frac{5}{6}y^{(4)} = -\frac{5}{16}y^{(0)}$$

$$y^{(8)} = -\frac{7}{8}y^{(6)} = \frac{35}{128}y^{(0)}$$

Varying n from 1 to ∞, skipping odd numbers, we get

$$y^{(n+2)} = -\frac{n+1}{n+2} y^{(n)}$$

$$y^{(3)} = -\frac{2}{3} y^{(1)}$$

$$y^{(5)} = -\frac{4}{5} y^{(3)} = \frac{8}{15} y^{(1)}$$

$$y^{(7)} = -\frac{6}{7} y^{(5)} = -\frac{16}{35} y^{(1)}$$

$$y^{(9)} = -\frac{8}{9} y^{(7)} = \frac{128}{315} y^{(1)}$$

(5-6.9)

Substituting by the derivatives of y from (5-6.8) and (6-6.9) in (5-4), with $\varepsilon = 0$, we get

$$y(x) = y^{(0)} + xy^{(1)} + \frac{x^2}{2!}\left(-\frac{1}{2} y^{(0)}\right) + \frac{x^3}{3!}\left(-\frac{2}{3} y^{(1)}\right) + \frac{x^4}{4!}\left(\frac{3}{8} y^{(0)}\right)$$
$$+ \frac{x^5}{5!}\left(\frac{8}{15} y^{(1)}\right) + \frac{x^6}{6!}\left(-\frac{5}{16} y^{(0)}\right) + \frac{x^7}{7!}\left(-\frac{16}{37} y^{(1)}\right) + \frac{x^8}{8!}\left(\frac{35}{128} y^{(0)}\right) + \frac{x^9}{9!}\left(\frac{128}{315} y^{(1)}\right)\ldots$$

(5-6-10)

Lumping terms in $y^{(0)}$ and $y^{(1)}$ together, we get

$$y(x) = y^{(0)}\left(1 - \frac{x^2}{2!}\left(\frac{1}{2}\right) + \frac{x^4}{4!}\left(\frac{3}{8}\right) - \frac{x^6}{6!}\left(\frac{5}{16}\right) + \frac{x^8}{8!}\left(\frac{35}{128}\right) + \ldots\right)$$
$$+ y^{(1)}\left(x - \frac{x^3}{3!}\left(-\frac{2}{3}\right) + \frac{x^5}{5!}\left(\frac{8}{15}\right) - \frac{x^7}{7!}\left(-\frac{16}{37}\right) + \frac{x^9}{9!}\left(\frac{128}{315}\right)\ldots\right)$$

(5-6-11)

The two arbitrary constants $y^{(0)}$ and $y^{(1)}$ are determined from the boundary conditions.

Example 83

Find a solution in power series in x of the equation

$$\frac{d^2 y(x)}{dx^2} + 3(x-1)\frac{dy(x)}{dx} - 6y = 0$$

(5-7.1)

In the neighborhood x = 1.

Solution

(a) Transform the independent variable by

$$x = t + 1 \tag{5-7.2a}$$

Thus, equation (5-7.1) becomes

$$\frac{d^2y}{dt^2} + 3t\frac{dy}{dt} - 6y = 0 \tag{5-7.2a}$$

(b) Second, construct **recurrence formula** for the coefficients of Taylor's series by differentiating (5-7.1) n times

Each term in equation (5-7.1) can be differentiated n times to give

$$\frac{d^n}{dx^n}(y'') = y^{(n+2)}$$
$$\frac{d^n}{dx^n}(3ty') = 3(ty^{(n+1)} + ny^{(n)}) \tag{5-7.3}$$
$$\frac{d^n}{dx^n}(-6y) = -6y^{(n)}$$

Where exponents denote order of derivatives.

Adding the above three and equating to zero, we get the (n+2)-order of equation (5-7.2)

$$y^{(n+2)} + 3ty^{(n+1)} + 3(ny^{(n)} - 2y^{(n)}) = 0 \tag{5-7.4}$$

The recurrence formula for the given boundary condition, x = 0 implies that t = 1 is therefore,

$$y^{(n+2)} = -3(n-2)y^{(n)} \tag{5-7.5}$$

Equation (5-7.5) gives the coefficients in Taylor's series

Similar to the previous example, we note that we have **two unknown constants, $y^{(0)}$ and $y^{(1)}$**, required to determine the final series expansion.

First, we need the initial value y at x = 0, which corresponds to $y^{(0)}$ in the above recurrence formula.
Second, we initial value of y' as x = 0,(t =1), since the recurrence formula in (5-7.5) skips one order of differentiation, jumps from (n-1) to (n+2).

Determination of the coefficients of Taylor's series

Starting from the recurrence formula (5-7.5), we get

$$y^{(n+2)} = -3(n-2)y^{(n)}$$
$$y^{(2)} = 6y^{(0)}$$
$$y^{(3)} = -3y^{(1)}$$
$$y^{(4)} = 0$$
$$y^{(5)} = -3y^{(3)} = 9y^{(1)}$$
$$y^{(6)} = 0$$
$$y^{(7)} == -3(3)y^{(5)} - 81y^{(1)}$$
$$y^{(8)} = 0$$
$$y^{(9)} = -3(5)y^{(7)} = 15(9)y^{(1)}$$

(5-7.6)

Substituting by the derivatives of y from (5-7.8) and (6-6.9) in (5-4), with $\varepsilon = 0$, we get

$$y(x) = y^{(0)} + ty^{(1)} + \frac{t^2}{2!}\left(6y^{(0)}\right) + \frac{t^3}{3!}\left(-3y^{(1)}\right) + \frac{t^5}{5!}\left(9y^{(1)}\right) + \frac{t^7}{7!}\left(-81y^{(1)}\right) + \frac{t^9}{9!}\left(135y^{(1)}\right)..$$

(5-7.7)

Lumping terms in $y^{(0)}$ and $y^{(1)}$ together, we get

$$y(x) = y^{(0)}\left(1 + 3t^2\right) + y^{(1)}\left(-\frac{3}{3!}t^3 + \frac{9}{5!}t^5 - \frac{81}{7!}t^7 + \frac{135}{9!}t^6 - ...\right)$$

(5-7.8)

Substituting by $t = x-1$, from (5-7.2a), we get

$$y(x) = y^{(0)}\left(1 + 3(x-1)^2\right) + y^{(1)}\left(-\frac{3}{3!}(x-1)^3 + \frac{9}{5!}(x-1)^5 - \frac{81}{7!}(x-1)^7 + \frac{135}{9!}(x-1)^6 - ...\right)$$

(5-7-8)

The two arbitrary constants $y^{(0)}$ and $y^{(1)}$ are determined from the boundary conditions. Thus, equation (5-7.8) comprises the Taylor's series solution of the o.d.e. (5-7.1) in the neighborhood of x = 0.

5-3. Solution by Frobenius's series expansion

Frobenius's method of solution applies to second-order o.d.e. with **variable coefficients** of the form

$$\frac{d^2y}{dx^2} + P(x)\frac{dy}{dx} + Q(x)y = 0$$

(5-8.1)

By expanding the variable coefficients P(x) and Q(x) in the neighborhood of x = 0 by the following sum of power series:

$$P(x) = \frac{1}{x}\sum_{n=0}^{\infty} A_n x^n \qquad (5.8.2)$$

$$Q(x) = \frac{1}{x^2}\sum_{n=0}^{\infty} B_n x^n \qquad (5.8.3)$$

(a) We can immediately notice that division by x in P and by x^2 in Q **homogenized the degree of power** on the independent variable x throughout the differential equation.

(b) We also immediately notice that at x = 0, P(x) may have infinite value only at **first order**.

(c) We also immediately notice that at x = 0, Q(x) may have infinite value only at **second order**.

(d) If P(x) has infinity of **first order** or less and Q(x) has infinity of **second order** or less at x = a (in lieu of x = 0), then the independent variable changes to **t = x –a**.

(e) Ordinary point of series expansions (5.8.2) and (5.8.3)

An **ordinary point** x =0 in the expansions (5-8.2) and (5-8.3) is defined as the point where P(x) and Q(x) converge at x = 0 to **finite values** if at least

$$A_0 = 0$$
$$B_0 = B_1 = 0 \qquad (5.8.4)$$

$$\lim_{x \to 0} P(x) = \lim_{x \to 0}\left(\frac{1}{x}\sum_{n=0}^{\infty} A_n x^n\right) \neq \infty \qquad (5.8.5)$$

$$\lim_{x \to 0} Q(x) = \lim_{x \to 0}\left(\frac{1}{x^2}\sum_{n=0}^{\infty} B_n x^n\right) \neq \infty \qquad (5.8.6)$$

(f) Regular point of expansions (5.8.2) and (5.8.3)

A **regular point** x =0 in the expansions (5.8.2) and (5.8.3) is defined as the point where xP(x) and $x^2Q(x)$ converge at x = 0 to **finite** as follows

$$\lim_{x \to 0} x P(x) = \lim_{x \to 0}\left(\sum_{n=0}^{\infty} A_n x^n\right) \neq \infty \qquad (5.8.7)$$

$$\lim_{x \to 0} x^2 Q(x) = \lim_{x \to 0}\left(\sum_{n=0}^{\infty} B_n x^n\right) \neq \infty \qquad (5.8.8)$$

(g) Singular point of expansions (5.8.2) and (5.8.3)

If the above P, Q, xP, and x^2Q do not converge at x = 0, then x = 0 is **singular point** for the series expansions and do not fit the Frobenius method.

Proof

Substitute by trial solution, in the form of expansion sum of power series, of equation (5-8.1) as follows:

$$y(x) = x^c \sum_{n=0}^{\infty} a_n x^n \qquad (5.9.1)$$

Differentiate twice to get

$$y'(x) = \sum_{n=0}^{\infty} a_n (n+c) x^{n+c-1}$$

$$y''(x) = \sum_{n=0}^{\infty} a_n (n+c)(n+c-1) x^{n+c-2} \qquad (5.9.2)$$

Substitute by the expansions of P(x), Q(x), y(x), and y'(x), and y" (x) in (5-8.1), we get

$$\left(\sum_{n=0}^{\infty} a_n (n+c)(n+c-1) x^{n+c-2} \right) + \left(\frac{1}{x} \sum_{n=0}^{\infty} A_n x^n \right) \left(\sum_{n=0}^{\infty} a_n (n+c) x^{n+c-1} \right)$$

$$+ \left(\frac{1}{x^2} \sum_{n=0}^{\infty} B_n x^n \right) \left(x^c \sum_{n=0}^{\infty} a_n x^n \right) = 0 \qquad (5-9.3)$$

Equating the coefficients of **equal powers** of x by zero (because the RHS is zero), we get:

(a) The **recurrence formula** for the coefficient a_n's.

(b) The possible value of the **exponent c** in equation (5.9.1) required to satisfy the differential equation.

The lowest power of x corresponds to n = 0. Thus, from equation (5-9.3), we get

$$a_0 c(c-1) + A_0 a_0 c + B_0 a_0 = 0 \qquad (5-9.4)$$

Therefore,

$$c^2 + c(A_0 - 1) + B_0 = 0 \tag{5-9.5a}$$

i.e., $$c = -\frac{1}{2}(A_0 - 1) \pm \frac{1}{2}\sqrt{(A_0 - 1)^2 - 4B_0} \tag{5-9.5b}$$

This is called the **indicial equation** as it identifies the possible expansion of $y(x)$ that satisfies equation (5-8.1).

The roots of (5-9.5b) determines the cases of solutions as follows.

5-3.1. Roots of indicial equation are distinct but differ by rational numbers

Example 84

Find a power series solution of the following equation by Frobenius's method

$$2x^2 \frac{d^2y(x)}{dx^2} - x\frac{dy(x)}{dx} + (1 - x^2)y = 0 \tag{5-10.1}$$

Solution

(a) Recurrence formula for coefficients of $y(x)$ sum power expansion

Arranging the terms the o.d.e. into terms of ascending weight as follows

$$\underbrace{\left[2x^2 \frac{d^2y(x)}{dx^2} - x\frac{dy(x)}{dx} + y\right]}_{\text{weight zero}} - \underbrace{\left[x^2 y\right]}_{\text{weight 2}} = 0 \tag{5-10.2}$$

Substituting by the series expansion from (5-9.1) into (5-10.2), we get

$$\left[2x^2D^2 - xD + 1\right]\sum_{n=0}^{\infty} a_n x^{n+c} - \left[\sum_{n=0}^{\infty} a_n x^{n+c+2}\right] = 0 \tag{5-10.3a}$$

Executing the differentiation, we get

$$\sum_{n=0}^{\infty} a_n \left[2(n+c)(n+c-1) - (n+c) + 1\right]x^{n+c} - \left[\sum_{n=0}^{\infty} a_n x^{n+c+2}\right] = 0 \tag{5-10.3b}$$

Hence, we have two terms of powers (n+c) and (n+c+2) of the independent variable x.

Thus, the recurrence formula is obtained by equating the coefficients of terms of equal power of x zero such that

$$a_n\left[2(n+c)(n+c-1) - (n+c) + 1\right] - a_{n-2} = 0 \tag{5-10.4a}$$

i.e.,

$$a_n = \frac{a_{n-2}}{(2(n+c)-1)(n+c-1)} \qquad (5\text{-}10.4b)$$

(b) The **indicial equation**

Substituting by $n = 0$ in equation (5-10.3b) and equating the coefficients of lowest power to zero, we get

$$[2c(c-1)-c+1]x^c - x^{c+2} = 0 \qquad (5\text{-}10.5)$$

Thus, the **identical equation** is obtained by equating the bracketed term to zero

$$2c^2 - 3c + 1 = 0 \qquad (5\text{-}10.6)$$

The two roots are

$$c_1 = 1, \qquad c_2 = \frac{1}{2} \qquad (5\text{-}10.7)$$

Similarly, putting $n = 1$, in (5-10.3b), we get

$$a_1|c + 2c^2|x^{1+c} - a_1 x^{1+c+2} = 0 \qquad (5\text{-}10.8)$$

Equating the coefficients of equal powers and noting that term containing c does not vanish for any value of c, therefore,

$$a_1 = 0 \qquad (5\text{-}10.9)$$

(c) Coefficients of series expansion of y(x)

The recurrence formula, (5-10.4b) and zero value of a_1, and the two values of c, we get

$$c_1 = 1$$
$$a_2 = \frac{a_0}{(2)(5)}$$
$$a_4 = \frac{a_2}{(9)(4)} = \frac{a_0}{(2)(5)(9)(4)} \qquad (5\text{-}10.11)$$
$$a_6 = \frac{a_4}{(13)(8)} = \frac{a_0}{(2)(5)(9)(4)(13)(8)}$$

Thus, the first solution, corresponding to $c_1 = 1$, is

$$y_1(x) = xa_0\left(1 + \frac{x^2}{(2)(5)} + \frac{x^4}{(2)(5)(9)(4)} + \frac{x^4}{(2)(5)(9)(4)(13)(8)} + \ldots\right) \qquad (5\text{-}10.12)$$

Similarly, second root $c = \frac{1}{2}$, gives

$$c_2 = \frac{1}{2}$$

$$a_2 = \frac{a_0}{(2)(3)}$$

$$a_4 = \frac{a_2}{(4)(7)} = \frac{a_0}{(2)(3)(4)(7)} \qquad (5\text{-}10.13)$$

$$a_6 = \frac{a_4}{(6)(11)} = \frac{a_0}{(2)(3)(4)(7)(6)(11)}$$

Thus, the second solution, corresponding to $c_2 = \frac{1}{2}$, is

$$y_2(x) = \sqrt{x}a_0\left(1 + \frac{x^2}{(2)(3)} + \frac{x^4}{(2)(3)(4)(7)} + \frac{x^4}{(2)(3)(4)(7)(6)(11)} + \ldots\right) \qquad (5\text{-}10.14)$$

The **general solution** comprises a linear combination of y_1 and y_2.

$$y(x) = y_1(x) + y_2(x)$$

$$= xA\left(1 + \frac{x^2}{(2)(5)} + \frac{x^4}{(2)(5)(9)(4)} + \frac{x^4}{(2)(5)(9)(4)(13)(8)} + \ldots\right) \qquad (5\text{-}10.15)$$

$$+ \sqrt{x}B\left(1 + \frac{x^2}{(2)(3)} + \frac{x^4}{(2)(3)(4)(7)} + \frac{x^4}{(2)(3)(4)(7)(6)(11)} + \ldots\right)$$

Where, A and B are constants

Example 85

Find a power series solution of the following equation by Frobenius's method

$$2x\frac{d^2y(x)}{dx^2} + (2x+1)\frac{dy(x)}{dx} + 3y = 0 \qquad (5\text{-}11.1)$$

Solution

(a) Recurrence formula for coefficients of y(x) sum power expansion

Arranging the terms the o.d.e. into terms of ascending weight as follows

$$\left[2x\frac{d^2y(x)}{dx^2} + \frac{dy(x)}{dx}\right] + \left[2x\frac{dy(x)}{dx} + 3y\right] = 0 \qquad (5\text{-}11.2)$$

$$\underbrace{\qquad}_{\text{weight -1}} \qquad \underbrace{\qquad}_{\text{weight zero}}$$

Substituting by the series expansion from (5-9.1) into (5-11.2), we get

$$\left[2xD^2 + D\right]\sum_{n=0}^{\infty} a_n x^{n+c} + \left[2xD + 3\right]\sum_{n=0}^{\infty} a_n x^{n+c} = 0 \qquad (5\text{-}11.3a)$$

Executing the differentiation, we get

$$\sum_{n=0}^{\infty} a_n\left[2(n+c)(n+c-1) + (n+c)\right]x^{n+c-1} + \sum_{n=0}^{\infty} a_n\left[2(n+c)+3\right]x^{n+c} = 0 \qquad (5\text{-}11.3b)$$

Hence, we have two terms of powers $(n+c)$ and $(n+c-1)$ of the independent variable x.

Thus, the recurrence formula is obtained by equating the coefficients of terms of equal power of x zero such that

$$a_n\left[(n+c)(2n+2c-1)\right] + a_{n-1}\left[(2(n+c-1)+3\right] = 0 \qquad (5\text{-}11.4a)$$

i.e.,

$$a_n = -a_{n-1}\frac{2n+2c+1}{(n+c)(2n+2c-1)} \qquad (5\text{-}11.4b)$$

(b) The **indicial equation**

Substituting by $n = 0$ in equation (5-11.3b) and equating the coefficients of lowest power to zero, we get

$$a_0\left[2c(c-1) + c\right]x^{c-1} + a_0\left[(2c+3)\right]x^c = 0 \qquad (5\text{-}11.5)$$

Thus, the **identical equation** is obtained by equating the left bracketed term to zero

$$2c^2 - c = 0 \qquad (5\text{-}11.6)$$

The two roots are

$$c_1 = 0, \qquad c_2 = \frac{1}{2} \tag{5-11.7}$$

(c) Coefficients of series expansion of y(x)

The recurrence formula, (5-11.4b) and the two values of c, we get

$$c_1 = 0$$
$$a_1 = -3a_0$$
$$a_2 = -a_1 \frac{5}{(2)(3)} = a_0 \frac{5}{(2)}$$
$$a_3 = -a_2 \frac{7}{(3)(5)} = -a_0 \frac{7}{(3)(2)} \tag{5-11.8}$$
$$a_4 = -a_3 \frac{9}{(4)(7)} = a_0 \frac{9}{(4)(3)(2)}$$

i.e.,

$$a_n = a_0 \frac{(-1)^n (2n+1)}{(n!)}$$

Thus, the first solution, corresponding to $c_1 = 0$, is

$$y_1(x) = a_0 \left(\sum_{n=0}^{\infty} (-1)^n \frac{(2n+1)}{(n!)} x^n \right) \tag{5-11.9}$$

Similarly, second root $c_2 = \frac{1}{2}$, gives

$$c_2 = \frac{1}{2}$$
$$a_1 = -a_0 \frac{4}{(3)}$$
$$a_2 = -a_1 \frac{6}{(5)(2)} = a_0 \frac{6(4)}{(5)(3)(2)} == a_0 \frac{(2^2)(2+1)}{(5)(3)}$$
$$a_3 = -a_2 \frac{8}{(7)(3)} = -a_0 \frac{8(6)(4)}{(7)(3)(5)(3)(2)} = -a_0 \frac{(2^3)(3+1)}{(7)(5)(3)} \tag{5-11.10}$$
$$a_4 = -a_3 \frac{10}{(9)(4)} = a_0 \frac{10(8)(6)(4)}{(9)(4)(7)(3)(5)(3)(2)} = a_0 \frac{(2^4)(4+1)}{(9)(4)(7)(5)(3)}$$

$$a_n = (-1)^n \frac{2^n (n+1)}{(2n+1)....(7)(5)(3)} a_0$$

Thus, the second solution, corresponding to $c = \frac{1}{2}$, is

$$y_2(x) = \sqrt{x} a_0 \sum_{n=0}^{\infty} (-1)^n \frac{2^n(n+1)}{(2n+1)....(7)(5)(3)} x^n \qquad (5\text{-}11.11)$$

The **general solution** comprises a linear combination of y_1 and y_2.

$$y(x) = y_1(x) + y_2(x)$$

$$= A \sum_{n=0}^{\infty} (-1)^n \frac{(2n+1)}{(n!)} x^n + \sqrt{x} B \sum_{n=0}^{\infty} (-1)^n \frac{2^n(n+1)}{(2n+1)....(7)(5)(3)} x^n \qquad (5\text{-}11.12)$$

Where, A and B are constants

5-3.2. Roots of indicial equation are equal

Example 86

Find the power series solution of the **Bessel's equation** by Frobenius's method

$$x \frac{d^2 y(x)}{dx^2} + \frac{dy(x)}{dx} + xy = 0 \qquad (5\text{-}12.1)$$

Solution

(a) Recurrence formula for coefficients of y(x) sum power expansion

Arranging the terms the o.d.e. into terms of ascending weight as follows

$$\left[x \frac{d^2 y(x)}{dx^2} + \frac{dy(x)}{dx} \right] + [xy] = 0 \qquad (5\text{-}12.2)$$
$$\text{weight -1} \qquad \text{weight +1}$$

Substituting by the series expansion from (5-9.1) into (5-12.2), we get

$$\left[xD^2 + D \right] \sum_{n=0}^{\infty} a_n x^{n+c} + \sum_{n=0}^{\infty} a_n x^{n+c+1} = 0 \qquad (5\text{-}12.3a)$$

Executing the differentiation, we get

$$\sum_{n=0}^{\infty} a_n \left[(n+c)(n+c-1) + (n+c) \right] x^{n+c-1} + \sum_{n=0}^{\infty} a_n x^{n+c+1} = 0 \qquad (5\text{-}12.3b)$$

Hence, we have two terms of powers (n+c+1) and (n+c-1) of the independent variable x.

Thus, the recurrence formula is obtained by equating the coefficients of terms of equal power of x zero such that

$$a_n(n+c)^2 + a_{n-2} = 0 \qquad (5\text{-}12.4a)$$

i.e.,

$$a_n = -\frac{a_{n-2}}{(n+c)^2} \qquad (5\text{-}12.4b)$$

(b) The indicial equation

Substituting by n = 0 in equation (5-12.3b) and equating the coefficients of lowest power to zero, we get

$$a_0[c(c-1)+c]x^{c-1} + a_0x^{c+1} = 0 \qquad (5\text{-}12.5)$$

Thus, the identical equation is obtained by equating the bracketed term to zero

$$c^2 = 0 \qquad (5\text{-}12.6)$$

The two roots are equal

$$c_1 = 0, \qquad c_2 = 0 \qquad (5\text{-}12.7)$$

Substituting by n = 1 in equation (5-12.3b) and equating the coefficients of equal power to zero, we get

$$\sum_{n=0}^{\infty} a_1[1]x^0 + \sum_{n=0}^{\infty} a_1x^2 = 0 \qquad (5\text{-}12.8a)$$

Therefore,

$$a_1 = 0 \qquad (5\text{-}12.8b)$$

(c) Coefficients of series expansion of y(x)

We will keep the roots of the indicial equations, c_1 and c_2 unknown in order to get another solution of the initial differential equation (5-12.1) as follows.

From, the recurrence formula, (5-12.4b) and the value of $a_1 = 0$, we get

210

$$a_1 = a_3 = a_5 = \ldots = 0$$

$$a_2 = -\frac{a_0}{(2+c)^2}$$

$$a_4 = -\frac{a_2}{(4+c)^2} = \frac{a_0}{(4+c)^2(2+c)^2} \tag{5-12.9}$$

$$a_6 = -\frac{a_4}{(6+c)^2} = -\frac{a_0}{(6+c)^2(4+c)^2(2+c)^2}$$

$$a_n = (-1)^{\frac{n}{2}}\frac{a_0}{(n+c)^2(n-2+c)^2\ldots(2+c)^2}$$

Thus, the first solution is

$$y_1(x) = a_0 x^c \left(1 - \frac{x^2}{(2+c)^2} + \frac{x^4}{(2+c)^2(4+c)^2} - \frac{x^6}{(2+c)^2(4+c)^2(6+c)^2} + \ldots\right) \tag{5-12.10}$$

(d) Finding the second particular solution when the two roots of indicial equations are equal

Since the two roots of equation (5-12.6) are equal, we will **circumvent such difficulty** by retaining c as unknown, substituting by $y_1(x)$, from (5-12.10) in the initial equation (5-12.6), as follows

$$x\frac{d^2 y(x)}{dx^2} + \frac{dy(x)}{dx} + xy = 0$$

$$a_0\left(c(c-1)x^{c-1} - \frac{(c+1)x^{c+1}}{(2+c)} + \frac{(c+3)x^{3+c}}{(2+c)^2(4+c)} - \frac{(c+5)x^{5+c}}{(2+c)^2(4+c)^2(6+c)} + \ldots\right) \tag{5-12.11}$$

$$+ a_0\left(cx^{c-1} - \frac{x^{1+c}}{(2+c)} + \frac{x^{3+c}}{(2+c)^2(4+c)} - \frac{x^{5+c}}{(2+c)^2(4+c)^2(6+c)} + \ldots\right)$$

$$+ a_0 x^{c+1}\left(1 - \frac{x^2}{(2+c)^2} + \frac{x^4}{(2+c)^2(4+c)^2} - \frac{x^6}{(2+c)^2(4+c)^2(6+c)^2} + \ldots\right) = 0$$

This can be arranged to eliminate canceling terms as follows

211

$$a_0 \left((c^2 - c)x^{c-1} - \frac{(c+1)x^{c+1}}{(2+c)} + \frac{(c+3)x^{3+c}}{(2+c)^2(4+c)} - \frac{(c+5)x^{5+c}}{(2+c)^2(4+c)^2(6+c)} + .. \right)$$

$$+ a_0 \left(cx^{c-1} \quad - \frac{x^{1+c}}{(2+c)} + \frac{x^{3+c}}{(2+c)^2(4+c)} - \frac{x^{5+c}}{(2+c)^2(4+c)^2(6+c)} + .. \right)$$

$$+ a_0 \left(\frac{(2+c)x^{c+1}}{(2+c)} - \frac{(4+c)x^{3+c}}{(2+c)^2(4+c)} + \frac{(6+c)x^{5+c}}{(2+c)^2(4+c)^2(6+c)} - .. \right) = 0 \qquad (5\text{-}12.12)$$

========== add the three equation by memebers ==================

$$a_0 \left(c^2 x^{c-1} - 0 + 0 - .. + .. \right) = 0$$

Before proceeding in proving that the **partial derivative of y_1 with respect to c** provides the second particular solution for the initial condition, we note that the above equation is the solution of the initial differential equation and can be written in the differential operator form as follows:

$$\left(xD^2 + D + x \right) y_1 = a_0 c^2 x^{c-1} = 0 \qquad (5\text{-}12.13)$$

As we expect, the above equation is satisfied by c = 0.

Let us differentiate the above equation with respect to c, as follows

$$\frac{\partial}{\partial c} \left(xD^2 + D + x \right) y_1 = \frac{\partial}{\partial c} a_0 c^2 x^{c-1} = 0 \qquad (5\text{-}12.14a)$$

i.e., $$\left(xD^2 + D + x \right) \frac{\partial y_1}{\partial c} = 2a_0 cx^{c-1} + a_0 c^2 \frac{\partial}{\partial c} \left(x^{c-1} \right) = 0 \qquad (5\text{-}12.14b)$$

Note the following differential substitution

$$x^{c-1} = \varphi$$
$$(c - 1)\ln x = \ln \varphi$$
$$\frac{\partial}{\partial c}\left[(c-1)\ln x \right] = \frac{\partial}{\partial c}\ln \varphi \qquad (5\text{-}12.14c)$$
$$\ln x = \frac{1}{\varphi}\frac{\partial \varphi}{\partial c}$$
$$\frac{\partial \varphi}{\partial c} = \varphi \ln x$$

Therefore,

$$\frac{\partial}{\partial c}\left(x^{c-1} \right) = \frac{\partial \varphi}{\partial c} = \varphi \ln x \qquad (5\text{-}12.14d)$$
$$= x^{c-1} \ln x$$

212

Therefore, equation (5-12-14a) becomes

$$2a_0cx^{c-1} + a_0c^2x^{c-1}\ln x = 0 \qquad (5\text{-}12.14e)$$

Thus, we could claim that we have another particular solution for $c = 0$ distinct from y_1, which also satisfy the initial differential equation.

(e) Partial derivative of first particular solution w.r.t. an arbitrary variable as a second particular solution of Bessel's o.d.e.

As has shown above, the second particular solution is obtained by differentiating y_1, equation (5-12.10), w.r.t. c as follows.

$$y_2(x) = \frac{\partial y_1(x)}{\partial c}\bigg|_{c=0}$$

$$= a_0\frac{\partial}{\partial c}x^c\left(1 - \frac{x^2}{(2+c)^2} + \frac{x^4}{(2+c)^2(4+c)^2} - \frac{x^6}{(2+c)^2(4+c)^2(6+c)^2} + ..\right) \qquad (5\text{-}12.15a)$$

Differentiating the composite function product gives

$$y_2(x) = a_0x^c\frac{\partial}{\partial c}\left(1 - \frac{x^2}{(2+c)^2} + \frac{x^4}{(2+c)^2(4+c)^2} - \frac{x^6}{(2+c)^2(4+c)^2(6+c)^2} + ..\right)$$

$$+ a_0\left(1 - \frac{x^2}{(2+c)^2} + \frac{x^4}{(2+c)^2(4+c)^2} - \frac{x^6}{(2+c)^2(4+c)^2(6+c)^2} + ..\right)\frac{\partial}{\partial c}x^c \qquad (5\text{-}12.15b)$$

Using the derivative of power exponent, equation (5-12.14c), we get

$$y_2(x) = a_0x^c\frac{\partial}{\partial c}\left(1 - \frac{x^2}{(2+c)^2} + \frac{x^4}{(2+c)^2(4+c)^2} - \frac{x^6}{(2+c)^2(4+c)^2(6+c)^2} + ..\right)$$

$$+ a_0x^c\ln x\left(1 - \frac{x^2}{(2+c)^2} + \frac{x^4}{(2+c)^2(4+c)^2} - \frac{x^6}{(2+c)^2(4+c)^2(6+c)^2} + ..\right) \qquad (5\text{-}12.15b)$$

(f) Determining the partial derivative w.r.t. c

The partial derivative in the above equation can be determined by the substitution

$$u(c,n) = \frac{1}{(2+c)^2(4+c)^2.......(2n+c)^2} \qquad (5\text{-}12.16a)$$

Where, n is the serial constant in the polynomial given above.

Taking the logarithm of both sides, we get

$$\ln u(c,n) = -2[\ln(2+c) + \ln(4+c) + \ldots + \ln(2n+c)] \qquad (5\text{-}12.16b)$$

Differentiating both sides w.r.t. c, we get

$$\frac{1}{u(c,n)} \frac{\partial u(c,n)}{\partial c} = -2\left[\frac{1}{(2+c)} + \frac{1}{(4+c)} + \ldots + \frac{1}{(2n+c)}\right]$$

Thus,

$$\frac{\partial u(c,n)}{\partial c} = -2u(c,n)\left[\frac{1}{(2+c)} + \frac{1}{(4+c)} + \ldots + \frac{1}{(2n+c)}\right] \qquad (5\text{-}12.16c)$$

Substituting by the expression of u(c,n) from equation (5-12.16a), and putting c = 0, we get

$$\left.\frac{\partial u(c,n)}{\partial c}\right|_{c=0} = -\frac{2}{(2)^2(4)^2\ldots(2n)^2}\left[\frac{1}{(2)} + \frac{1}{(4)} + \ldots + \frac{1}{(2n)}\right]$$
$$= -\frac{1}{(2)^2(4)^2\ldots(2n)^2}\left[1 + \frac{1}{2} + \frac{1}{3}\ldots + \frac{1}{n}\right] \qquad (5\text{-}12.16d)$$

(g) Final expression for second particular solution

Before substituting by the above derivative of u(c,n) in equation (5-12.15b), we need to arrange the terms in the latter equation to show the location of the obtained derivative, as follows.

$$y_2(x) = a_0 x^c \frac{\partial}{\partial c}\left(\sum_{n=1}^{\infty}\frac{(-1)^n x^{2n}}{(2+c)^2(4+c)^2\ldots(2n+c)^2}\right) + y_1(x)\ln x \qquad (5\text{-}12.17a)$$

Substituting by u(c,n) and c = 0, from (5-12.16a), we get

$$y_2(x) = a_0 \frac{\partial}{\partial c}\left(\sum_{n=1}^{\infty}(-1)^n x^{2n} u(c,n)\right) + y_1(x)\ln x$$
$$= a_0\left(\sum_{n=1}^{\infty}(-1)^n\left[x^{2n}\frac{\partial u(c,n)}{\partial c}\right]\right) + y_1(x)\ln x \qquad (5\text{-}12.17b)$$

Therefore, the final form of y_2 is obtained by substituting from (5-12.16d) into (5-12.17b), we get

$$y_2(x) = a_0\left(\sum_{n=1}^{\infty}\left[\frac{(-1)^{n+1} x^{2n}}{(2)^2(4)^2\ldots(2n)^2}\left(1 + \frac{1}{2} + \frac{1}{3}\ldots + \frac{1}{n}\right)\right]\right) + y_1(x)\ln x \quad (5\text{-}12.17c)$$

214

(h) General Solution of Bessel's equation

$$y(x) = Ay_1(x) + By_2(x)$$

$$= \left[\begin{array}{l} A(1 + \ln x)\left(1 - \frac{x^2}{2^2} + \frac{x^4}{2^2 4^2} - \frac{x^6}{2^2 4^2 6^2} + ..\right) \\ \\ + B\left(\frac{x^2}{2^2} - \frac{x^4}{(2)^2(4)^2}\left(1 + \frac{1}{2}\right) + \frac{x^6}{(2)^2(4)^2(6)^2}\left(1 + \frac{1}{2} + \frac{1}{3}\right) - ..\right) \end{array} \right] \qquad (5\text{-}12.18a)$$

This can be farther simplified to

$$y(x) = Ay_1(x) + By_2(x)$$

$$= A(1 + \ln x)\left(1 - \frac{x^2}{2^2} + \frac{x^4}{2^4(2!)^2} - \frac{x^6}{2^6(3!)^2} + ..\right) + B\left(\frac{x^2}{2^2} - \frac{x^4\left(1 + \frac{1}{2}\right)}{2^4(2!)^2} + \frac{x^6\left(1 + \frac{1}{2} + \frac{1}{3}\right)}{2^6(3!)^2} - ..\right)$$

$$(5\text{-}12.18b)$$

(i) Conclusion

When the two roots of the indicial equation are equal, there exist two particular solutions, one obtained from the value of the two roots in equation (5.9.1), the other by differentiating the first particular solution with respect to the exponent c, equation (5-12.15a).

Example 87

Find the series solution of the equation by Frobenius's method

$$x\frac{d^2y(x)}{dx^2} + (x+1)\frac{dy(x)}{dx} + 2y = 0 \qquad (5\text{-}13.1)$$

Solution

(a) Recurrence formula for coefficients of y(x) sum power expansion

Arranging the terms the o.d.e. into terms of ascending weight as follows

$$\underbrace{\left[x\frac{d^2y(x)}{dx^2} + \frac{dy(x)}{dx}\right]}_{\text{weight -1}} + \underbrace{\left[x\frac{dy(x)}{dx} + 2y\right]}_{\text{weight 0}} = 0 \qquad (5\text{-}13.2)$$

Substituting by the series expansion from (5-9.1) into (5-13.2), we get

$$\left[xD^2 + D\right]\sum_{n=0}^{\infty} a_n x^{n+c} + \left[xD + 2\right]\sum_{n=0}^{\infty} a_n x^{n+c} = 0 \qquad (5\text{-}13.3a)$$

Executing the differentiation, we get

$$\sum_{n=0}^{\infty} a_n\left[(n+c)(n+c-1) + (n+c)\right]x^{n+c-1} + \sum_{n=0}^{\infty} a_n\left[(n+c) + 2\right]x^{n+c} = 0 \qquad (5\text{-}13.3b)$$

Hence, we have two terms of powers (n+c+1) and (n+c) of the independent variable x.

Thus, the recurrence formula is obtained by equating the coefficients of terms of equal power of x zero such that

$$a_n(n+c)^2 + a_{n-1}(n+c+1) = 0 \qquad (5\text{-}13.4a)$$

i.e.,

$$a_n = -\frac{a_{n-1}(n+c+1)}{(n+c)^2} \qquad (5\text{-}13.4b)$$

(b) The indicial equation

Substituting by $n = 0$ in equation (5-13.3b) and equating the coefficients of lowest power to zero, we get

$$a_0 c^2 x^{c-1} + a_0(c+2)x^{c+1} = 0 \qquad (5\text{-}13.5)$$

Thus, the **identical equation** is obtained by equating the left term (lowest power in x) to zero

$$c^2 = 0 \qquad (5\text{-}13.6)$$

The two roots are equal

$$c_1 = 0, \qquad c_2 = 0 \qquad (5\text{-}13.7)$$

We need not find a1 because the recurrence formula (5-13.4b) **does not skip** coefficients.

(c) Coefficients of series expansion of y(x)

We will keep the roots of the indicial equations, c_1 and c_2 **unknown** in order to get another solution of the initial differential equation (5-13.1) as follows.

216

From, the recurrence formula, (5-13.4b), we get

$$a_1 = -\frac{a_0(c+2)}{(c+1)^2}$$

$$a_2 = a_0\frac{(c+3)}{(c+2)(c+1)^2}$$

$$a_3 = -a_0\frac{(c+4)}{(c+3)(c+2)(c+1)^2}$$ (5-13.8)

$$a_4 = a_0\frac{(c+5)}{(c+4)(c+3)(c+2)(c+1)^2}$$

$$\dots\dots\dots\dots\dots\dots\dots$$

$$a_n = a_0\frac{(-1)^n(c+n+1)}{(c+n)..(c+3)(c+2)(c+1)^2}$$

Thus, the first solution is

$$y_1(x) = a_0 x^c\left(1 + \sum_{n=1}^{\infty}\frac{(-1)^n(c+n+1)x^n}{(c+n)..(c+3)(c+2)(c+1)^2}\right)$$ (5-13.9)

(d) Partial derivative of first particular solution w.r.t. an arbitrary variable as a second particular solution of Bessel's o.d.e.

Similar to equation (5-12.15a), we get

$$y_2(x) = \frac{\partial y_1(x)}{\partial c}$$

$$= a_0 x^c\frac{\partial}{\partial c}\left(1 + \sum_{n=1}^{\infty}\frac{(-1)^n(c+n+1)x^n}{(c+n)..(c+3)(c+2)(c+1)^2}\right) + y_1(x)\ln x$$ (5-13.10b)

(e) Determining the partial derivative w.r.t. c

The partial derivative in the above equation can be determined by the substitution

$$u(c,n) = \frac{(c+n+1)}{(c+n)..(c+3)(c+2)(c+1)^2}$$ (5-13.11a)

Where, n is the serial constant in the polynomial given above.

Taking the logarithm of both sides, we get

$$\ln u(c,n) = \ln(c+n+1) - [\ln(c+n)+..+\ln(c+3)+\ln(c+2)+2\ln(c+1)]$$ (5-13.11b)

217

Differentiating both sides w.r.t. c, we get

$$\frac{1}{u(c,n)}\frac{\partial u(c,n)}{\partial c} = \frac{1}{c+n+1} - \left[\frac{1}{(c+n)} + .. + \frac{1}{(c+3)} + \frac{1}{(c+2)} + \frac{2}{(c+1)}\right]$$

Thus,

$$\frac{\partial u(c,n)}{\partial c} = u(c,n)\left(\frac{1}{c+n+1} - \left[\frac{1}{(c+n)} + .. + \frac{1}{(c+3)} + \frac{1}{(c+2)} + \frac{2}{(c+1)}\right]\right) \quad (5\text{-}13.11c)$$

Substituting by the expression of u(c,n) from equation (5-13.11a), and putting c = 0, we get

$$\left.\frac{\partial u(c,n)}{\partial c}\right|_{c=0} = \frac{n+1}{n!}\left(\frac{1}{n+1} - \left[\frac{1}{n} + .. + \frac{1}{3} + \frac{1}{2} + \frac{2}{1}\right]\right) \quad (5\text{-}13.11d)$$

(g) Final expression for second particular solution

Therefore, the final form of y_2 is obtained by substituting from (5-13.11d) into (5-13.10b), we get

$$y_2(x) = \frac{\partial y_1(x)}{\partial c}$$

$$= a_0 \sum_{n=1}^{\infty} x^n \frac{\partial}{\partial c}\left(\frac{(-1)^n(c+n+1)}{(c+n)..(c+3)(c+2)(c+1)^2}\right) + y_1(x)\ln x \quad (5\text{-}13.12a)$$

$$= a_0 \sum_{n=1}^{\infty} x^n (-1)^n \frac{n+1}{n!}\left(\frac{1}{n+1} - \left[\frac{1}{n} + .. + \frac{1}{3} + \frac{1}{2} + \frac{2}{1}\right]\right) + y_1(x)\ln x$$

Where, y_1 is obtained from (5-13.9) by putting c = 0

$$y_1(x) = a_0 x^c\left(1 + \sum_{n=1}^{\infty} \frac{(-1)^n(n+1)x^n}{n!}\right) \quad (5\text{-}13.12b)$$

(h) General Solution of Bessel's equation

$$y(x) = Ay_1(x) + By_2(x)$$

$$= A(1+\ln x)\left(1 + \sum_{n=1}^{\infty} \frac{(-1)^n(n+1)x^n}{n!}\right) + B\sum_{n=1}^{\infty} x^n(-1)^n \frac{n+1}{n!}\left(\frac{1}{n+1} - \left[\frac{1}{n} + .. + \frac{1}{3} + \frac{1}{2} + \frac{2}{1}\right]\right)$$

$$(5\text{-}13.12c)$$

5-3.3. Roots of indicial equation differing by integer yield one finite solution

One of the roots of the indicial equation leads to coefficients of infinite values.

Example 88

Find the power series solution of the **Bessel's equation** by Frobenius's method

$$x^2 \frac{d^2y(x)}{dx^2} + x\frac{dy(x)}{dx} + (x^2 - 1)y = 0 \qquad (5\text{-}14.1)$$

Solution

(a) Recurrence formula for coefficients of y(x) sum power expansion

Arranging the terms the o.d.e. into terms of ascending weight as follows

$$\left[x^2 \frac{d^2y(x)}{dx^2} + x\frac{dy(x)}{dx} - y\right] + \left[x^2 y\right] = 0 \qquad (5\text{-}14.2)$$
$$\underset{\text{weight -0}}{} \qquad \underset{\text{weight + 2}}{}$$

Substituting by the series expansion from (5-9.1) into (5-12.2), we get

$$\left[x^2 D^2 + xD - 1\right]\sum_{n=0}^{\infty} a_n x^{n+c} + \sum_{n=0}^{\infty} a_n x^{n+c+2} = 0 \qquad (5\text{-}14.3a)$$

Executing the differentiation, we get

$$\sum_{n=0}^{\infty} a_n \left[(n+c)(n+c-1) + (n+c) - 1\right]x^{n+c} + \sum_{n=0}^{\infty} a_n x^{n+c+2} = 0 \qquad (5\text{-}14.3b)$$

Hence, we have two terms of powers $(n+c+2)$ and $(n+c)$ of the independent variable x.

Thus, the recurrence formula is obtained by equating the coefficients of terms of equal power of x zero such that

$$a_n(n+c-1)(n+c+1) + a_{n-2} = 0 \qquad (5\text{-}14.4a)$$

i.e.,

$$a_n = -\frac{a_{n-2}}{(n+c+1)(n+c-1)} \qquad (5\text{-}14.4b)$$

219

(b) The **indicial equation**

Substituting by $n = 0$ in equation (5-14.3b) and equating the coefficients of lowest power to zero, we get

$$a_0(c-1)(c+1)x^c + a_0 x^{c+2} = 0 \tag{5-14.5}$$

Thus, the **identical equation** is obtained by equating the bracketed term to zero

$$(c-1)(c+1) = 0 \tag{5-14.6}$$

The two roots are different by integer 2

$$c_1 = 1, \qquad c_2 = -1 \tag{5-14.7}$$

Substituting by $n = 1$ in equation (5-14.3b) and equating the coefficients of equal power to zero, we get

$$\sum_{n=0}^{\infty} a_1 c(c+2)x^{c+1} + \sum_{n=0}^{\infty} a_1 x^{c+3} = 0 \tag{5-14.8a}$$

Therefore, since the term containing c does not vanish for any value of c, we get

$$a_1 = 0 \tag{5-14.8b}$$

(c) **Coefficients of series expansion of y(x)**

We will keep the roots of the indicial equations, c_1 and c_2 **unknown** in order to get another solution of the initial differential equation (5-14.1) as follows.

From, the recurrence formula, (5-14.4b) and the value of $a_1 = 0$, we get

$$
\begin{aligned}
a_1 &= a_3 = a_5 = ... = 0 \\
a_2 &= -\frac{a_0}{(c+3)(c+1)} \\
a_4 &= \frac{a_0}{(c+5)(c+3)^2(c+1)} \\
a_6 &= -\frac{a_0}{(c+7)(c+5)^2(c+3)^2(c+1)}
\end{aligned}
\tag{5-14.9}
$$

Thus, the first solution is

$$y_1(x) = a_0 x^c \left(1 - \frac{x^2}{(c+3)(c+1)} + \frac{x^4}{(c+5)(c+3)^2(c+1)} - \frac{x^6}{(c+7)(c+5)^2(c+3)^2(c+1)} + ..\right) \quad (5\text{-}14.10)$$

Clearly, one of the roots, $c_1 = 1$, yields finite solution, while the other, $c_2 = -1$, leads to infinite coefficients as denominator $(c+1)$ is common to all coefficients in equation (5-14.9).

(d) Finding the second particular solution when the one root of indicial equations causes infinite coefficients

Substitute by

$$a_0 = k(c+1) \quad (5\text{-}14.11)$$

Where, k is substituted in y_1, equation (5-14.10), which in turn is substituted in the differential equation (5-14.1) to give

$$y_1(x) = k(c+1)x^c \left(1 - \frac{x^2}{(c+3)(c+1)} + \frac{x^4}{(c+5)(c+3)^2(c+1)} - \frac{x^6}{(c+7)(c+5)^2(c+3)^2(c+1)} + ..\right)$$

$$(5\text{-}14.12a)$$

and

$$\left[x^2\frac{d^2y(x)}{dx^2} + x\frac{dy(x)}{dx} - y\right] + \left[x^2 y\right] = 0$$

$$= kx^c\left((c+1)c(c-1) - \frac{(c+2)(c+1)x^2}{(c+3)} + \frac{(c+4)(c+3)x^4}{(c+5)(c+3)^2} - \frac{(c+6)(c+5)x^6}{(c+7)(c+5)^2(c+3)^2} + ..\right)$$

$$+ kx^c\left((c+1)c - \frac{(c+2)x^2}{(c+3)} + \frac{(c+4)x^4}{(c+5)(c+3)^2} - \frac{(c+6)x^6}{(c+7)(c+5)^2(c+3)^2} + ..\right) \quad (5\text{-}14.12b)$$

$$- kx^c\left((c+1) - \frac{x^2}{(c+3)} + \frac{x^4}{(c+5)(c+3)^2} - \frac{x^6}{(c+7)(c+5)^2(c+3)^2} + ..\right)$$

$$+ kx^c\left(\frac{(c+3)(c+1)x^2}{(c+3)} - \frac{x^4}{(c+3)} + \frac{x^6}{(c+5)(c+3)^2} - \frac{x^8}{(c+7)(c+5)^2(c+3)^2} + ..\right)$$

Arranging and canceling, we get

$$\left[x^2\frac{d^2y(x)}{dx^2} + x\frac{dy(x)}{dx} - y\right] + \left[x^2 y\right] = 0$$

$$(5\text{-}14.12c)$$

$$= kx^c(c+1)c^2 - kx^c(c+1)$$

This is what remains after canceling equal terms with opposite signs.

As before, we note that the substitution by knew constant $k(c+1)$, in lieu of a_0, in above equation is the solution of the initial differential equation and can be written in the differential operator form as follows:

$$(x^2D^2 + xD - 1)y_1 + x^2y_1 = kx^c(c+1)c^2 - kx^c(c+1)$$

$$= 0$$

(5-14.13)

As we expect, the above equation is satisfied by $c = -1$.

Therefore, the first particular solution is obtained from (5-14.12a) by substituting by $c = -1$.

$$y_1(x) = \frac{k}{x}\left(-\frac{x^2}{2} + \frac{x^4}{(4)(2)^2} - \frac{x^6}{(6)(4)^2(2)^2} + ..\right)$$

(5-14.14a)

Test for substituting by $c =1$ in equation (5-14.10)

$$y_1(x) = a_0x^c\left(1 - \frac{x^2}{(c+3)(c+1)} + \frac{x^4}{(c+5)(c+3)^2(c+1)} - \frac{x^6}{(c+7)(c+5)^2(c+3)^2(c+1)} + ..\right)$$

$$= a_0x\left(1 - \frac{x^2}{(4)(2)} + \frac{x^4}{(6)(4)^2(2)} - \frac{x^6}{(8)(6)^2(4)^2(2)} + ..\right)$$

(5-14.14b)

We note that $c = 1$ yields a multiple of the same solution in (5-14.14a), hence $c =1$ does not comprise a third particular solution.

(e) Partial derivative of first particular solution w.r.t. an arbitrary variable as a second particular solution of Bessel's o.d.e.

As has shown above, the second particular solution is obtained by differentiating y_1, equation (5-14.12a), w.r.t. c as follows.

$$y_2(x) = \frac{\partial y_1(x)}{\partial c}\bigg|_{c=0}$$

$$= k\frac{\partial}{\partial c}\left((c+1)x^c\left(1 - \frac{x^2}{(c+3)(c+1)} + \frac{x^4}{(c+5)(c+3)^2(c+1)} - \frac{x^6}{(c+7)(c+5)^2(c+3)^2(c+1)} + ..\right)\right)$$

(5-14.15a)

Differentiating the composite function product gives

222

$$y_2(x) = k \left[\begin{array}{l} \left[(c+1) - \dfrac{x^2}{(c+3)} + \dfrac{x^4}{(c+5)(c+3)^2} - \dfrac{x^6}{(c+7)(c+5)^2(c+3)^2} + .. \right] \dfrac{\partial}{\partial c}(x^c) \\ + x^c \dfrac{\partial}{\partial c}\left[(c+1) - \dfrac{x^2}{(c+3)} + \dfrac{x^4}{(c+5)(c+3)^2} - \dfrac{x^6}{(c+7)(c+5)^2(c+3)^2} + .. \right] \end{array} \right]$$

(5-14.15b)

Using the derivative of power exponent, equation (5-12.14c), we get

$$y_2(x) = kx^c \left[\begin{array}{l} \left[(c+1) - \dfrac{x^2}{(c+3)} + \dfrac{x^4}{(c+5)(c+3)^2} - \dfrac{x^6}{(c+7)(c+5)^2(c+3)^2} + .. \right] \ln x \\ + \dfrac{\partial}{\partial c}\left[(c+1) - \dfrac{x^2}{(c+3)} + \dfrac{x^4}{(c+5)(c+3)^2} - \dfrac{x^6}{(c+7)(c+5)^2(c+3)^2} + .. \right] \end{array} \right]$$

(5-14.15c)

(f) Determining the partial derivative w.r.t. c

The partial derivative in the above equation can be determined by the substitution

$$\frac{\partial}{\partial c}\left((c+1) - \frac{x^2}{(c+3)} + \frac{x^4}{(c+5)(c+3)^2} - \frac{x^6}{(c+7)(c+5)^2(c+3)^2} + .. \right)$$

$$= 1 + \frac{x^2}{(c+3)^2} + x^4\left(-2(c+5)^{-1}(c+3)^{-3} - (c+5)^{-2}(c+3)^{-2} \right)$$

$$- x^6\left[-(c+7)^{-2}(c+5)^{-2}(c+3)^{-2} - 2(c+7)^{-1}(c+5)^{-3}(c+3)^{-2} - 2(c+7)^{-1}(c+5)^{-2}(c+3)^{-3} \right] + ...$$

(5-14.16a)

Substituting by c = -1, we get

$$\frac{\partial}{\partial c}\left((c+1) - \frac{x^2}{(c+3)} + \frac{x^4}{(c+5)(c+3)^2} - \frac{x^6}{(c+7)(c+5)^2(c+3)^2} + .. \right)$$

$$= 1 - \frac{x^2}{(2)^2} + x^4\left(\frac{2}{(4)^1(2)^3} + \frac{1}{(4)^2(2)^2} \right) + x^6\left[\frac{1}{(6)^2(4)^2(2)^2} + \frac{2}{(6)^1(4)^3(2)^2} + \frac{2}{(6)^1(4)^2(2)^3} \right] + ...$$

(5-14.16b)

Therefore, the second particular solution, (5-14.15c) becomes

223

$$y_2(x) = k\left[\begin{array}{l}\left[\left(-\dfrac{x}{(2)} + \dfrac{x^3}{(4)(2)^2} - \dfrac{x^5}{(6)(4)^2(2)^2} + ..\right)\ln x\right. \\[4mm] + \dfrac{1}{x} - \dfrac{x}{(2)^2} + x^3\left(\dfrac{2}{(4)^1(2)^3} + \dfrac{1}{(4)^2(2)^2}\right) + x^4\left[\dfrac{1}{(6)^2(4)^2(2)^2} + \dfrac{2}{(6)^1(4)^3(2)^2} + \dfrac{2}{(6)^1(4)^2(2)^3}\right] + \end{array}\right]$$

(5-14.17)

(h) General Solution of Bessel's equation

$$y(x) = Ay_1(x) + By_2(x)$$

$$= A(1 + \ln x)\left(-\dfrac{x}{2} + \dfrac{x^3}{(4)(2)^2} - \dfrac{x^5}{(6)(4)^2(2)^2} + ...\right)$$

$$+ B\left[\begin{array}{l}\dfrac{1}{x} - \dfrac{x}{(2)^2} + x^3\left(\dfrac{2}{(4)^1(2)^3} + \dfrac{1}{(4)^2(2)^2}\right) \\[4mm] + x^5\left[\dfrac{1}{(6)^2(4)^2(2)^2} + \dfrac{2}{(6)^1(4)^3(2)^2} + \dfrac{2}{(6)^1(4)^2(2)^3}\right] + ...\end{array}\right]$$

(5-14.18)

(i) Conclusion

When the two roots of the indicial equation are equal but differing by an integer, the constant a_0 should be replaced by $k(c+\alpha)$ so as to remove the singularity from the recurrent formula for the coefficients of the series expansion.

Example 89

Find power series solution the equation by Frobenius's method

$$x\dfrac{d^2y(x)}{dx^2} + x\dfrac{dy(x)}{dx} + y = 0$$

(5-15.1)

Solution

(a) Recurrence formula for coefficients of y(x) sum power expansion

Arranging the terms the o.d.e. into terms of ascending weight as follows

$$\left[x\dfrac{d^2y(x)}{dx^2}\right] + \left[x\dfrac{dy(x)}{dx} + y\right] = 0$$

weight -1   weight 0

(5-15.2)

224

Substituting by the series expansion from (5-9.1) into (5-12.2), we get

$$\left[xD^2\sum_{n=0}^{\infty}a_nx^{n+c} + [xD+1]\sum_{n=0}^{\infty}a_nx^{n+c}\right] = 0 \qquad (5\text{-}15.3a)$$

Executing the differentiation, we get

$$\sum_{n=0}^{\infty}a_n(n+c)(n+c-1)x^{n+c-1} + \sum_{n=0}^{\infty}a_n(n+c+1)x^{n+c} = 0 \qquad (5\text{-}15.3b)$$

Hence, we have two terms of powers (n+c-1) and (n+c) of the independent variable x.

Thus, the recurrence formula is obtained by equating the coefficients of terms of equal power of x zero such that

$$a_n(n+c)(n+c-1) + a_{n-1}(n+c) = 0 \qquad (5\text{-}15.4a)$$

i.e.,

$$a_n = -\frac{a_{n-1}}{(n+c-1)} \qquad (5\text{-}15.4b)$$

(b) The **indicial equation**

Substituting by $n = 0$ in equation (5-15.3b) and equating the coefficients of lowest power to zero, we get

$$a_0c(c-1)x^{c-1} + a_0(c+1)x^c = 0 \qquad (5\text{-}15.5)$$

Thus, the **identical equation** is obtained by equating the coefficient of the term of lowest power to zero

$$c(c-1) = 0 \qquad (5\text{-}15.6)$$

The two roots are different by integer

$$c_1 = 0, \qquad c_2 = 1 \qquad (5\text{-}15.7)$$

We do not need to find a_1 since the recurrence formula does not skip indices.

(c) Coefficients of series expansion of y(x)

From, the recurrence formula, (5-15.4b), we get

225

$$a_1 = -\frac{a_0}{c}$$

$$a_2 = -\frac{a_1}{(c+1)} = \frac{a_0}{c(c+1)}$$

$$a_3 = -\frac{a_2}{(c+2)} = -\frac{a_0}{c(c+1)(c+2)} \qquad (5\text{-}15.8)$$

$$a_4 = -\frac{a_3}{(c+3)} = \frac{a_0}{c(c+1)(c+2)(c+3)}$$

$$a_n = (-1)^n \frac{a_0}{(n+c-1)....(c+2)(c+1)c}$$

Thus, the first solution is

$$y_1(x) = a_0 x^c \left(1 - \sum_{n=1}^{\infty} (-1)^n \frac{a_0 x^n}{(n+c-1)....(c+2)(c+1)c} \right) \qquad (5\text{-}15.9)$$

Clearly, one of the roots, $c_1 = 0$, yields singular solutions.

(d) Finding the second particular solution when the one root of indicial equations causes infinite coefficients

Therefore, substitute by

$$a_0 = kc \qquad (5\text{-}15.10)$$

Where, k is substituted in y_1, equation (5-15.10), which in turn is substituted in the differential equation (5-15.1) to give

$$y_1(x) = kx^c \left(c - x + \sum_{n=2}^{\infty} (-1)^n \frac{x^n}{(n+c-1)....(c+2)(c+1)} \right) \qquad (5\text{-}15.11)$$

(e) Partial derivative of first particular solution w.r.t. an arbitrary variable as a second particular solution of Bessel's o.d.e.

As has shown above, the second particular solution is obtained by differentiating y_1, equation (5-15.11), w.r.t. c as follows.

$$y_2(x) = \frac{\partial y_1(x)}{\partial c}\bigg|_{c=0}$$

$$= k \frac{\partial}{\partial c}\left(cx^c - x^{c+1} + \sum_{n=2}^{\infty} (-1)^n \frac{x^{n+c}}{(n+c-1)....(c+2)(c+1)} \right) \qquad (5\text{-}15.12)$$

$$= k\left(x^c\left(1 + (1-x)c\ln x\right) + \frac{\partial}{\partial c}\sum_{n=2}^{\infty} (-1)^n \frac{x^{n+c}}{(n+c-1)....(c+2)(c+1)} \right)$$

(f) Determining the partial derivative w.r.t. c

The partial derivative in the above equation can be determined by the substitution

$$u(c,n) = \frac{1}{(n+c-1)..(c+3)(c+2)(c+1)} \qquad (5\text{-}15.13a)$$

Where, n is the serial constant in the polynomial given above.

Taking the logarithm of both sides, we get

$$\ln u(c,n) = -\left[\ln(n+c-1) + .. + \ln(c+3) + \ln(c+2) + \ln(c+1)\right] \qquad (5\text{-}15.13b)$$

Differentiating both sides w.r.t. c, we get

$$\frac{1}{u(c,n)}\frac{\partial u(c,n)}{\partial c} = -\left[\frac{1}{(c+n-1)} + .. + \frac{1}{(c+3)} + \frac{1}{(c+2)} + \frac{1}{(c+1)}\right]$$

Thus,

$$\frac{\partial u(c,n)}{\partial c} = -u(c,n)\left[\frac{1}{(c+n-1)} + .. + \frac{1}{(c+3)} + \frac{1}{(c+2)} + \frac{1}{(c+1)}\right] \qquad (5\text{-}15.13c)$$

Substituting by the expression of u(c,n) from equation (5-15.13a), and putting c = 0, we get

$$\frac{\partial u(c,n)}{\partial c}\bigg|_{c=0} = \frac{1}{(n-1)}\left(-\left[\frac{1}{n-1} + .. + \frac{1}{3} + \frac{1}{2} + 1\right]\right) \qquad (5\text{-}15.13d)$$

Returning to the second particular solution, equation (5-15.12), to substitute by the derivative of u(c,n), we get

$$y_2(x) = k \left(\begin{array}{l} 1 + \ln x \sum_{n=2}^{\infty} (-1)^n \dfrac{x^{n+c}}{(n-1)!} \\ - \sum_{n=2}^{\infty} (-1)^n \dfrac{x^{n+c}}{(n-1)!} \left[\dfrac{1}{n-1} + .. + \dfrac{1}{3} + \dfrac{1}{2} + 1 \right] \end{array} \right)$$

(5-15.14)

(g) General Solution

Substitute by $c = 0$ in equation (5-15.11), we get the first particular solution

$$y_1(x) = k \left(-x + \sum_{n=2}^{\infty} (-1)^n \dfrac{x^n}{(n-1)!} \right)$$

(5-15.15)

Thus, from (5-15.14) and (5-15.15), the general solution of (5-15.1) becomes

$$y(x) = Ay_1(x) + By_2(x)$$
$$= A \left(-x + \sum_{n=2}^{\infty} (-1)^n \dfrac{x^n}{(n-1)!} \right)$$
$$+ B \left(\begin{array}{l} 1 + \ln x \sum_{n=2}^{\infty} (-1)^n \dfrac{x^{n+c}}{(n-1)!} \\ - \sum_{n=2}^{\infty} (-1)^n \dfrac{x^{n+c}}{(n-1)!} \left[\dfrac{1}{n-1} + .. + \dfrac{1}{3} + \dfrac{1}{2} + 1 \right] \end{array} \right)$$

(5-15.16)

5-3.4. Roots of indicial equation differing by integer but coefficients of power series become indeterminate

Example 90

Find the power series solution of the equation by Frobenius's method

$$(1 - x^2) \dfrac{d^2 y(x)}{dx^2} + 2x \dfrac{dy(x)}{dx} - 3y = 0$$

(5-16.1)

Solution

(a) Recurrence formula for coefficients of y(x) sum power expansion

Arranging the terms the o.d.e. into terms of ascending weight as follows

228

$$\left[-x^2\frac{d^2y(x)}{dx^2}+2x\frac{dy(x)}{dx}-3y\right]+\left[\frac{d^2y(x)}{dx^2}\right]=0 \qquad (5\text{-}16.2)$$

$$\underset{\text{weight -0}}{} \qquad \underset{\text{weight}-2}{}$$

Substituting by the series expansion from (5-9.1) into (5-12.2), we get

$$\left[-x^2D^2+2xD-3\right]\sum_{n=0}^{\infty}a_nx^{n+c}+\left[D^2\right]\sum_{n=0}^{\infty}a_nx^{n+c}=0 \qquad (5\text{-}16.3a)$$

Executing the differentiation, we get

$$\sum_{n=0}^{\infty}a_n\left[-(n+c)(n+c-1)+2(n+c)-3\right]x^{n+c}+\sum_{n=0}^{\infty}a_n(n+c)(n+c-1)x^{n+c-2}=0 \qquad (5\text{-}16.3b)$$

Hence, we have two terms of powers (n+c) and (n+c-2) of the independent variable x.

Thus, the recurrence formula is obtained by equating the coefficients of terms of **equal power of x** zero such that

$$a_n\left[(n+c)(-n-c+3)-3\right]+a_{n+2}(n+c+2)(n+c+1)=0 \qquad (5\text{-}16.4a)$$

i.e.,

$$a_{n+2}=a_n\frac{(n+c)(n+c-3)+3}{(n+c+2)(n+c+1)} \qquad (5\text{-}16.4b)$$

(b) The indicial equation

Substituting by n = 0 in equation (5-16.3b) and equating the coefficients of lowest power to zero, we get

$$a_0\left[-c(c-1)+2c-3\right]x^c+a_0c(c-1)x^{c+2}=0 \qquad (5\text{-}16.5)$$

Thus, the **identical equation** is obtained by term of lowest power to zero

$$c(c-1)=0 \qquad (5\text{-}16.6)$$

The two roots are different by an integer.

$$c_1=0, \qquad c_2=1 \qquad (5\text{-}16.7)$$

Substituting by n = 1 in equation (5-16.3b) and equating the coefficients of lowest power to zero, we get

$$\sum_{n=0}^{\infty} a_1\left[-(c+1)c + 2(c+1) - 3\right]x^c + \sum_{n=1}^{\infty} a_n(c+1)cx^{c-1} = 0 \qquad (5\text{-}16.8a)$$

Therefore, coefficient of lowest power in x is

$$a_1(c+1)c = 0 \qquad (5\text{-}16.8b)$$

Since, when c vanishes, a_1 becomes **indeterminate**. When c is equal to 1, a_1 vanishes.

(c) Two series expansion of y(x)

From, the recurrence formula, (5-16.4b), we get

$$a_2 = a_0 \frac{c(c-3)+3}{(c+2)(c+1)}$$

$$a_4 = a_2 \frac{(c+2)(c-1)+3}{(c+4)(c+3)} = a_0 \frac{[(c+2)(c-1)+3](c(c-3)+3)}{(c+4)(c+3)(c+2)(c+1)} \qquad (5\text{-}16.9a)$$

$$a_6 = a_4 \frac{(c+4)(c+1)+3}{(c+6)(c+4)} = a_0 \frac{[(c+4)(c+1)+3][(c+2)(c-1)+3](c(c-3)+3)}{(c+6)(c+5)(c+4)(c+3)(c+2)(c+1)}$$

....

$$a_3 = a_1 \frac{(c+1)(c-2)+3}{(c+3)(c+2)}$$

$$a_5 = a_3 \frac{(c+3)c+3}{(c+5)(c+4)} = a_1 \frac{[(c+3)c+3][(c+1)(c-2)+3]}{(c+5)(c+4)(c+3)(c+2)} \qquad (5\text{-}16.9b)$$

$$a_7 = a_5 \frac{(c+5)(c+2)+3}{(c+7)(c+6)} = a_1 \frac{[(c+5)(c+2)+3][(c+3)c+3][(c+1)(c-2)+3]}{(c+7)(c+6)(c+5)(c+4)(c+3)(c+2)}$$

....

Thus, when **c = 0 and a_1 is indeterminate**, we have two series expansions, one from terms of **even** degrees of x and one from terms of **odd** degrees of x, as follows.

$$y_1(x)_{even} = a_0 x^c \left(\begin{array}{l} 1 + x^2 \dfrac{(c(c-3)+3)}{(c+2)(c+1)} + x^4 \dfrac{[(c+2)(c-1)+3](c(c-3)+3)}{(c+4)(c+3)(c+2)(c+1)} \\ + x^6 \dfrac{[(c+4)(c+1)+3][(c+2)(c-1)+3](c(c-3)+3)}{(c+6)(c+5)(c+4)(c+3)(c+2)(c+1)} + ... \end{array} \right)$$

$$(5\text{-}16.10a)$$

And

$$y_1(x)_{odd} = a_1 x^c \left(\begin{array}{l} x + x^3 \dfrac{(c+1)(c-2)+3}{(c+3)(c+2)} + x^5 \dfrac{[(c+3)c+3][(c+1)(c-2)+3]}{(c+5)(c+4)(c+3)(c+2)} \\ + x^7 \dfrac{[(c+5)(c+2)+3][(c+3)c+3][(c+1)(c-2)+3]}{(c+7)(c+6)(c+5)(c+4)(c+3)(c+2)} + \dots \end{array} \right)$$

(5-16.10b)

If $c = 1$ and a_1 vanishes (i.e. no longer indeterminate).

Therefore, the two solutions are obtained by substituting by $c = 0$ and $c = 1$ in (5-16.10a) we get

$$y_1(x) = [y_1(x)_{even}]_{c=0}$$
$$= a_0\left(1 + x^2 \frac{3}{2} + x^4 \cdot \frac{3}{2} \cdot \frac{1}{12} + x^6 \cdot \frac{3}{2} \cdot \frac{1}{12} \cdot \frac{7}{30} + \dots\right)$$

(5-16.11a)

And

$$y_2(x) = [y_1(x)_{even}]_{c=1}$$
$$= a_1 x\left(1 + x^2 \frac{1}{6} + x^4 \cdot \frac{1}{6} \cdot \frac{3}{20} + x^6 \cdot \frac{1}{6} \cdot \frac{3}{20} \cdot \frac{13}{42} + \dots\right)$$

(5-16.11b)

(d) Conclusion

When the two roots of the indicial equation are differing by an integer, and one of the coefficients is indeterminate, then series with finite coefficient contains two constants.

Example 91

Find the power series solution of the equation by Frobenius's method

$$x \frac{d^2 y(x)}{dx^2} + (x+2)\frac{dy(x)}{dx} + y = 0$$

(5-17.1)

Solution

(a) Recurrence formula for coefficients of y(x) sum power expansion

Arranging the terms the o.d.e. into terms of ascending weight as follows

$$\left[x\frac{d^2 y(x)}{dx^2} + 2\frac{dy(x)}{dx}\right] + \left[x\frac{dy(x)}{dx} + y\right] = 0$$

$$\quad\text{weight -1}\qquad\qquad\text{weight} - 0$$

(5-17.2)

231

Substituting by the series expansion from (5-9.1) into (5-12.2), we get

$$\left[xD^2 + 2D\right]\sum_{n=0}^{\infty} a_n x^{n+c-1} + \left[xD + 1\right]\sum_{n=0}^{\infty} a_n x^{n+c} = 0 \tag{5-17.3a}$$

Executing the differentiation, we get

$$\sum_{n=0}^{\infty} a_n\left[(n+c)(n+c-1) + 2(n+c)\right]x^{n+c-1} + \sum_{n=0}^{\infty} a_n\left[n+c+1\right]x^{n+c} = 0 \tag{5-17.3b}$$

Hence, we have two terms of powers $(n+c-1)$ and $(n+c)$ of the independent variable x.

Thus, the recurrence formula is obtained by equating the coefficients of terms of **equal power of x** zero such that

$$a_n\left[(n+c)(n+c-1) + 2(n+c)\right] + a_{n-1}(n+c) = 0 \tag{5-17.4a}$$

i.e.,

$$a_n = -a_{n-1}\frac{1}{(n+c+1)} \tag{5-17.4b}$$

(b) The indicial equation

Substituting by $n = 0$ in equation (5-17.3b) and equating the coefficients of lowest power to zero, we get

$$a_0\left[c(c-1) + 2c\right]x^{c-1} + a_0(c+1)x^c = 0 \tag{5-17.5}$$

Thus, the **identical equation** is obtained by term of lowest power to zero

$$c(c+1) = 0 \tag{5-17.6}$$

The two roots are different by an integer.

$$c_1 = 0, \qquad c_2 = -1 \tag{5-17.7}$$

Substituting by $n = 1$ in equation (5-17.3b) and equating the coefficients of lowest power to zero, we get

$$\sum_{n=0}^{\infty} a_1\left[(c+1)c + 2(c+1)\right]x^c + \sum_{n=1}^{\infty} a_1(c+2)cx^{c+1} = 0 \tag{5-17.8a}$$

232

Therefore, coefficient of lowest power in x is

$$a_1(c+1)(c+2) = 0 \qquad (5\text{-}17.8b)$$

Since, when $c = -1$, a_1 becomes **indeterminate**. When c is equal to 0, a_1 vanishes.

(c) Two series expansion of y(x)

From, the recurrence formula, (5-17.4b), with $c = -1$, we get

$$a_2 = -a_1 \frac{1}{2}$$

$$a_3 = -a_2 \frac{1}{3} = a_1 \frac{1}{3} \cdot \frac{1}{2}$$

$$a_4 = -a_3 \frac{1}{4} = -a_1 \frac{1}{4} \frac{1}{3} \frac{1}{2} \qquad (5\text{-}17.9)$$

$$\dots\dots\dots\dots\dots\dots$$

$$a_n = a_1 \frac{(-1)^{n+1}}{n!}$$

Therefore, the two solutions are obtained by substituting by $c = -1$ in the series expansion

$$y_1(x) = x^{-1}\left(a_0 + a_1 \sum_{n=1}^{\infty} \frac{(-1)^{n+1}}{n!} x^n \right) \qquad (5\text{-}17.10)$$

This has the two constants and is written as the general solution as follows

$$y_1(x) = \frac{A}{x} + B\left(\sum_{n=1}^{\infty} \frac{(-1)^{n+1}}{n!} x^{n-1} \right) \qquad (5\text{-}17.11)$$

Similar to the previous example, substituting by $c = 0$ does not yield new solution.

5-4. Solution by reciprocal power series for asymptotes

For large values of the independent variable x, the o.d.e. with variable coefficients, equation (5-8.1) is modified by changing the form of the independent variable by a reciprocal variable such that

$$x = \frac{1}{t}, \qquad t = \frac{1}{x} \tag{5-18.1}$$

And the derivatives are

$$\frac{dt}{dx} = -\frac{1}{x^2} = -t^2$$

$$\frac{dy}{dx} = \frac{dy}{dt}\frac{dt}{dx} = -t^2\frac{dy}{dt} \tag{5-18.2}$$

$$\frac{d^2y}{dx^2} = t^4\frac{d^2y}{dt^2} + 2t^3\frac{dy}{dt}$$

Substituting by the above derivatives in equation (5-8.1), we get

$$t^4\frac{d^2y}{dt^2} + 2t^3\frac{dy}{dt} - t^2\frac{dy}{dt}P(t) + Q(t)y = 0 \tag{5-18.3}$$

i.e., $\quad t^4\frac{d^2y}{dt^2} + \left(2t^3 - t^2P(t)\right)\frac{dy}{dt} + Q(t)y = 0$

The reciprocal differential equation is solved by the power series as above.

5-5. Exercises on solution of second-order linear o.d.e. with variable coefficients by power series

Solve the following o.d.e. using Taylor and Frobenius methods

5-1.
$$x\frac{d^2y}{dx^2} + (x+a)\frac{dy}{dx} + ay = 0$$

In the cases:
(a) $a = \frac{1}{2}$,
(b) 0,
(c) 1,
(d) 2

5-2.
$$x\frac{d^2y}{dx^2} + (x^2+1)\frac{dy}{dx} + xy = 0$$

5-3.
$$x^2\frac{d^2y}{dx^2} + x\frac{dy}{dx} + (x^2-n^2)y = 0$$

In the cases:
(a)   n is not an integer
(b)   n is an integer

5-4.
$$\frac{d^2y}{dx^2} + 3x\frac{dy}{dx} - 6y = 0$$

5-5.
$$\frac{d^2y}{dx^2} + 4x\frac{dy}{dx} - 4y = 0$$

5-6.
$$(x^2+1)\frac{d^2y}{dx^2} - y = 0$$

5-7.
$$(x^2+9)\frac{d^2y}{dx^2} + y = 0$$

5-8.
$$(1-x^2)\frac{d^2y}{dx^2} - 2x\frac{dy}{dx} + m(m+1)y = 0$$

Given that m is a positive integer:

(a) m is even
(b) m is odd

5-9. $$\frac{d^2y}{dx^2} - 2x\frac{dy}{dx} + 2my = 0$$

Given that m is a positive integer:
(a) m is even
(b) m is odd

5-10. $$x\frac{d^2y}{dx^2} + (1-x)\frac{dy}{dx} + my = 0$$

Given that m is a positive integer:
(a) m is even
(b) m is odd

5-11. $$\frac{d^2y}{dx^2} - 2x\frac{dy}{dx} - 4y = 0$$

5-12. $$x^2\frac{d^2y}{dx^2} + x\frac{dy}{dx} + (x^2 - \frac{1}{4})y = 0$$

5-13. $$(x^2 - x)\frac{d^2y}{dx^2} + (2x+1)\frac{dy}{dx} - \frac{1}{x}y = 0$$

5-14. $$x\frac{d^2y}{dx^2} + \frac{1}{2}\frac{dy}{dx} + y = 0$$

5-15. $$(x^3 + 2x^2)\frac{d^2y}{dx^2} + (3x^2 - x)\frac{dy}{dx} + (x+1)y = 0$$

5-16. $$\frac{d^2y}{dx^2} + \frac{1}{2x}\frac{dy}{dx} + y = 0$$

5-17. $$(2x - x^2)\frac{d^2y}{dx^2} + (1 - 4x)\frac{dy}{dx} - 2y = 0$$

5-18. $(1 - x^2)\dfrac{d^2 y}{dx^2} - x\dfrac{dy}{dx} = 0$

5-19. $(1 + x^2)\dfrac{d^2 y}{dx^2} + 2x\dfrac{dy}{dx} = 0$

CHAPTER 6

PARTIAL DIFFERENTIAL EQUATION

6-1. Definition of partial differential equation

1. Dependent functions depend **on more than one independent variable**

e.g.,
$$y = f(x, z, t, ...)$$
$$\frac{\partial y}{\partial x} = \frac{\partial}{\partial x} f(x, z, t, ...)$$
(6-1.1)

2. The **order** of the differential equation is the order of the higher derivative

e.g.,
$$\frac{\partial^4 y}{\partial x^4} + \frac{\partial y}{\partial x} = g(x, z, t, ...) \text{ is a fourth-order partial diff. eq.}$$
(6-1.2)

3. **Wave propagation** of light

e.g.,
$$\left(\frac{\partial u}{\partial x}\right)^2 + \left(\frac{\partial u}{\partial y}\right)^2 + \left(\frac{\partial u}{\partial z}\right)^2 = n(x, y, z)$$
(6-1.3)

4. **Heat equation**

e.g.,
$$\frac{\partial u}{\partial t} = a^2 \frac{\partial^2 u}{\partial x^2}$$
(6-1.4)

5. **Vibration of string**

e.g.,
$$\frac{\partial^2 u}{\partial t^2} = a^2 \frac{\partial^2 u}{\partial x^2}$$
(6-1.5)

6. **Laplace's equation**

e.g.,
$$\frac{\partial^2 u}{\partial x^2} + \frac{\partial^2 u}{\partial y^2} + \frac{\partial^2 u}{\partial z^2} = 0$$
(6-1.6)

6-2. Classification of partial differential equations

6-2.a. Linear p.d.e.

Linear partial differential equation of the second-order is defined as

$$A(x,y)\frac{\partial^2 \varphi(x,y)}{\partial x^2} + B(x,y)\frac{\partial^2 \varphi(x,y)}{\partial x \partial x} + C(x,y)\frac{\partial^2 \varphi(x,y)}{\partial y^2} = 0 \qquad (6-2.1)$$

1. This equation is linear because the **coefficients of its derivatives** are devoid of the dependent function φ and only functions of the independent variables, x and y.

2. Linearly independent solutions of linear p.d.e. can be **combined linearly** to comprise a solution of the same p.d.e.

3. Linear combinations of particular solutions facilitate satisfying **complex boundary conditions** that cannot be satisfied by individual solutions.

4. Linear combinations of particular solution facilitate devising **basic** and **correcting tensors** such that one tensor describes general distribution of a function and the correcting tensor accounts of imposed boundary conditions.

The following example shows the concept of linear combination of linearly independent solutions:

$$\begin{pmatrix} \sigma_x & \tau_{xy} & \tau_{xz} \\ \tau_{yx} & \sigma_y & \tau_{yz} \\ \tau_{zx} & \tau_{zy} & \sigma_z \end{pmatrix} = \begin{pmatrix} \sigma_{x,0} & \tau_{xy,0} & \tau_{xz,0} \\ \tau_{yx,0} & \sigma_{y,0} & \tau_{yz,0} \\ \tau_{zx,0} & \tau_{zy,0} & \sigma_{z,0} \end{pmatrix} + \sum_{m=1}^{\infty} a_m \begin{pmatrix} \sigma_{x,m} & \tau_{xy,m} & \tau_{xz,m} \\ \tau_{yx,m} & \sigma_{y,m} & \tau_{yz,m} \\ \tau_{zx,m} & \tau_{zy,m} & \sigma_{z,m} \end{pmatrix} \qquad (6-2.2)$$

general tesnor   basic tensor   correcting tensor

This is a common tensor description of normal σ and shear stresses τ in elastic material. The basic tensor describes symmetric stresses in some principal plane in the bulk of matter. The correcting tensor accounts for deviation from the principal plane. The summation of linearly independent solutions accounts for complex geometries and properties of matter.

5. The **quadratic characteristic equation** of (6-2.1) can be written in terms that facilitates classification as follows

$$Q(x,y) = A(x,y)U^2 + B(x,y)UV + C(x,y)V^2 = 0 \qquad (6-2.3)$$

6-2.1. Elliptic p.d.e. $\{[B(x,y)]^2 - 4A(x,y)C(x,y)\} < 0$

1. Provided that A(x,y) does not vanish, that the criteria $(B^2-4AC < 0)$ is satisfied at all points in the region.

2. Q(x,y) is positive for all values of U and V.

3. Q(x,y) can be transformed to the form

$$\frac{\partial^2 \psi(x,y)}{\partial X^2} + \frac{\partial^2 \psi(x,y)}{\partial Y^2} = Q\left(X, Y, \frac{\partial \psi(x,y)}{\partial X}, \frac{\partial \psi(x,y)}{\partial Y}\right) \tag{6-3.1}$$

Where $X(x,y)$ and $Y(x,y)$ are transformation function.

4. Example

$$\nabla^2 u = \frac{\partial^2 u}{\partial x^2} + \frac{\partial^2 u}{\partial y^2} + \frac{\partial^2 u}{\partial z^2} = 0 \tag{6-3.2}$$

6-2.2. Parabolic p.d.e. $\{[B(x,y)]^2 - 4A(x,y)C(x,y)\} = 0$

1. Provided that $(B^2-4AC = 0)$ is satisfied at all points in the region.

2. Q(x,y) vanishes.

3. Q(x,y) can be transformed to the form

$$\frac{\partial^2 \psi(x,y)}{\partial Y^2} = Q\left(X, Y, \psi, \frac{\partial \psi(x,y)}{\partial X}, \frac{\partial \psi(x,y)}{\partial Y}\right) \tag{6-3.3}$$

4. Example

$$\frac{\partial^2 T}{\partial x^2} = \frac{1}{a}\frac{\partial T}{\partial t} = 0 \tag{6-3.4}$$

6-2.3. Hyperbolic p.d.e. $\{[B(x,y)]^2 - 4A(x,y)C(x,y)\} > 0$

1. Provided that $(B^2-4AC > 0)$ is satisfied at all points in the region.

2. Q(x,y) is negative at all points in the region.

3. Q(x,y) can be transformed to the form

240

$$\frac{\partial^2 \psi(x,y)}{\partial X \partial Y} = Q\left(X, Y, \psi, \frac{\partial \psi(x,y)}{\partial X}, \frac{\partial \psi(x,y)}{\partial Y}\right)$$ (6-3.5)

6-2.b. Nonlinear p.d.e.

1. Various particular solutions cannot be combined linearly and any combination of particular solutions does not comprise a solution of the initial p.d.e.

2. We will **not** deal with nonlinear p.d.e. in this book.

3. Example: Burgers' Equation

$$\frac{\partial \varphi(x,t)}{\partial t} + \varphi(x,t)\frac{\partial \varphi(x,t)}{\partial x} = \mu\frac{\partial^2 \varphi(x,t)}{\partial x^2}$$ (6-4)

This is a gas diffusion equation that entails the dependent function multiplied by its derivative, which precludes linear combinations of solutions in terms of independent variables.

6-3. Separation of variables or geometrical polynomials

The approach of choosing algebraic polynomial for the quadratic function Q(x,y) reduces the choice to single coordinates via the products of polynomials, or trigonometric polynomials.

Example 92: Airy-Lévy's equation

Solve the **Airy-Lévy's equation** by geometric polynomials of two separate variables

$$\frac{\partial^4 \varphi(x,y)}{\partial x^4} + 2\frac{\partial^4 \varphi(x,y)}{\partial y^2 \partial x^2} + \frac{\partial^4 \varphi(x,y)}{\partial y^4} = 0$$ (6-5.1)

Solution

Consider the solution

$$\varphi(x,y) = X(x)Y(y)$$ (6-5.2)

Differentiate the above solution with respect to each of the independent variables x and y and substitute in the initial equation (6-5.1) to get

$$Y(y)\frac{d^4X(x)}{dx^4} + 2\frac{d^2X(x)}{dx^2} \cdot \frac{d^2Y(y)}{dy^2} + X(x)\frac{d^4Y(y)}{dy^4} = 0 \qquad (6\text{-}5.3)$$

(ii) Choice of periodic polynomial functions

In order to disentangle the X and Y in equation (6-5.3), we will choose periodic functions that allow us to assume the following solutions

$$\frac{d^4X(x)}{dx^4} = k^4X(x) \qquad (6\text{-}5.4a)$$

$$\frac{d^2X(x)}{dx^2} = -k^2X(x) \qquad (6\text{-}5.4b)$$

The periodicity of X is proven by differentiating (6-5.4) twice to get

$$\frac{d^4X(x)}{dx^4} = -k^2\frac{d^2X(x)}{dx^2} \qquad (6\text{-}6.3)$$

From (6-5.4) and (6-6.3), we discern the periodicity since we have

$$\frac{d^4X(x)}{dx^4} = -k^2\frac{d^2X(x)}{dx^2} = -k^2\left(-k^2X(x)\right) = k^4X(x) \qquad (6\text{-}6.4)$$

Thus, the choice of the negative sign after every two differentiations guaranteed the periodicity of X(x).

Now, substituting from equations (6-5.4) and (6-5.4) into (6-5.3) we get

$$Y(y)k^4X(x) - 2k^2X(x) \cdot \frac{d^2Y(y)}{dy^2} + X(x)\frac{d^4Y(y)}{dy^4} = 0 \qquad (6\text{-}5.5)$$

$$\left(Y(y)k^4 - 2k^2 \cdot \frac{d^2Y(y)}{dy^2} + \frac{d^4Y(y)}{dy^4}\right)X(x) = 0 \qquad (6\text{-}5.6)$$

Thus, from equations (6-5.6) and (6-5.4) we can write two separate differential equations for X(x) and Y(y) as follows:

$$\frac{d^2X(x)}{dx^2} + k^2X(x) = 0 \qquad (6\text{-}5.7)$$

$$Y(y)k^4 - 2k^2 \cdot \frac{d^2Y(y)}{dy^2} + \frac{d^4Y(y)}{dy^4} = 0 \qquad (6\text{-}5.8)$$

242

Thus, we obtained two separate o.d.e. that could be integrated along out previous described methods as follows

$$\frac{d^2 X(x)}{dx^2} + k^2 X(x) = 0 \qquad\qquad (6\text{-}5.7)$$

$$Y(y)k^4 - 2k^2 \cdot \frac{d^2 Y(y)}{dy^2} + \frac{d^4 Y(y)}{dy^4} = 0 \qquad\qquad (6\text{-}5.8)$$

$$\left(D^2 + k^2\right)X(x) = 0$$

$$\qquad\qquad (6\text{-}5.8a)$$

$$\left(D^4 - 2D^2 k^2 + k^4\right)Y(y) = 0 \qquad\qquad (6\text{-}5.8b)$$

In those two equations, we have used the differential D-operator to mean differentiation with respect to the independent variable of each equation separate from the other.

The characteristic equations are

$$m^2 + k^2 = 0$$

$$\qquad\qquad (6\text{-}5.9a)$$

$$n^4 - 2n^2 k^2 + k^4 = 0$$

$$\left(n - k\right)^2 \left(n + k\right)^2 = 0 \qquad\qquad (6\text{-}5.9b)$$

The two imaginary roots of (6-5.9a) give the solution

$$X(x) = A\cos kx + B\sin kx \qquad\qquad (6\text{-}5.10a)$$

The four roots of equation (6-5.10a) comprise a pair of roots, each of two equal roots, giving the solution

$$Y(y) = C_1 \cosh ky + C_2 y \cosh ky + C_3 \sinh ky + C_4 y \sinh ky \qquad\qquad (6\text{-}5.10b)$$

Therefore, the solution of equation (6-5.1), given by (6-5.2) is obtained by multiplying (6-5.10a) and (6-5.10b) to give

$$\varphi(x,y) = X(x)Y(y)$$

$$\qquad\qquad (6\text{-}5.11)$$

$$= \left(A\cos kx + B\sin kx\right)\!\left[C_1 \cosh ky + C_2 y \cosh ky + C_3 \sinh ky + C_4 y \sinh ky\right]$$

Example 93: Elastic Vibration

A vibration of deflection u(x,t) at distance x, measured from a fixed end of a stretched homogeneous membrane, is described by the equation

$$\frac{\partial^2 u}{\partial t^2} = a^2 \frac{\partial^2 u}{\partial x^2}$$

(6-6.1a)

Given the boundary conditions:
$$u(0, t) = u(l, t) = 0,$$ for all t   (6-6.1b)

And initial conditions

$$u(x,0) = 3\sin\frac{2\pi}{l}x$$
$$\frac{\partial u(x,0)}{\partial t} = 0$$ for all $0 \le x \le l$ (6-6.1c)

Find the deflection function at any time t and distance x.

Solution

(a) Separation of variables

Consider the solution

$$u(x, t) = X(x)T(t)$$

(6-6.2)

Differentiate the above solution with respect to each of the independent variables x and t and substitute in the initial equation (6-6.1) to get

$$X\frac{d^2 T}{dt^2} = a^2 T \frac{d^2 X}{dx^2}$$

(6-6.3a)

Divide both sides by $a^2 XT$, we get

$$\frac{1}{a^2 T}\frac{d^2 T}{dt^2} = \frac{1}{X}\frac{d^2 X}{dx^2}$$

(6-6.3b)

Since each side of the above equation is independent of the other side in regard to its variable, we have

$$\frac{1}{a^2 T}\frac{d^2 T}{dt^2} = \frac{1}{X}\frac{d^2 X}{dx^2} = -k^2$$

(6-6.4)

Where k is an arbitrary constant of proportionality and the negative sign is suggested to ease solution without affecting the outcome.

We could then separate the two sides of the above equations into two independent equations as follows

$$\frac{d^2T}{dt^2} + k^2a^2T = 0$$

$$\frac{d^2X}{dx^2} + k^2X = 0$$

(6-6.5)

The two ordinary differential equation immediately yield the solutions

$$T(t) = A\cos kat + B\sin kat$$
$$X(x) = C\cos kx + F\sin kx$$

(6-6.6)

Thus, the **particular solution**, (6-6.2) becomes

$$u(x, t) = X(x)T(t)$$
$$= (A\cos kat + B\sin kat)[C\cos kx + F\sin kx]$$

(6-6.7)

(b) Boundary and Initial conditions

Boundary conditions

The four constants A, B, C, and F are determined from the four boundary and initial conditions given by equations (6-6.1b and 1c), as follows.

Equation (6-6.1b), at $x = 0$ and $x = l$, gives upon substituting in equation (6-6.7)

$$0 = (A\cos kat + B\sin kat)[C]$$

(6-6.8a)

Thus, C must vanish since the term containing functions in t cannot vanish at all times.

At $x = l$, we get

$$0 = (A\cos kat + B\sin kat)[F\sin kl]$$

(6-6.8b)

The roots of the sine term implies

$$kl = 0, \pm\pi, \pm n\pi$$
$$k = \frac{n\pi}{l}, \qquad n = 0,1,2,...$$

(6-6.8c)

Thus, after satisfying the boundary conditions, we have

$$u(x,t) = \left(A\cos\frac{n\pi at}{l} + B\sin\frac{n\pi at}{l} \right)\left[\sin\frac{n\pi}{l}x \right] \qquad (6\text{-}6.8d)$$

Initial conditions

Equation (6-6.1c), at t = 0 and all x, gives upon substituting in equation (6-6.8d) and its derivative the following

$$u(x,0) = \sum_{n=1}^{\infty} A_n\left[\sin\frac{n\pi}{l}x \right] = 3\sin\frac{2\pi}{l}x$$

$$\frac{\partial u(x,0)}{\partial t} = \left(\frac{n\pi a}{l}B \right)\left[\sin\frac{n\pi}{l}x \right] = 0 \qquad (6\text{-}6.9a)$$

Therefore, B must vanish at all x and A is given by

$$A_2 = 3$$
$$B = 0 \qquad (6\text{-}6.9a)$$
$$n = 2$$

Since all values of A_n vanish if n not equal to 2

$$u(x,t) = 3\cos\frac{2\pi at}{l}\sin\frac{2\pi}{l}x \qquad (6\text{-}6.9b)$$

Example 94: Heat Equation

Given the heat equation

e.g.,
$$\frac{\partial u}{\partial t} = a^2\frac{\partial^2 u}{\partial x^2} \qquad (6\text{-}7.1a)$$

Given the boundary conditions:

$$u(0,t) = u(l,t) = 0, \qquad\qquad \text{for all t} \qquad (6\text{-}7.1b)$$

And initial conditions

$$u(x,0) = x, \quad \text{for} \quad 0 \le x \le l$$
$$\frac{\partial u(x,\infty)}{\partial t} = \text{finite} \qquad (6\text{-}7.1c)$$

246

Find the u(x,t) at any time t and distance x.

Solution

(a) Separation of variables

Consider the solution

$$u(x,t) = X(x)T(t) \tag{6-7.2}$$

Differentiate the above solution with respect to each of the independent variables x and t and substitute in the initial equation (6-7.1) to get

$$X\frac{dT}{dt} = a^2T\frac{d^2X}{dx^2} \tag{6-7.3a}$$

Divide both sides by a^2XT, we get

$$\frac{1}{a^2T}\frac{dT}{dt} = \frac{1}{X}\frac{d^2X}{dx^2} \tag{6-7.3b}$$

Since each side of the above equation is independent of the other side in regard to its variable, we have

$$\frac{1}{a^2T}\frac{dT}{dt} = \frac{1}{X}\frac{d^2X}{dx^2} = -k^2 \tag{6-7.4}$$

Where k is an arbitrary constant of proportionality and the negative sign is suggested to ease solution without affecting the outcome.

We could then separate the two sides of the above equations into two independent equations as follows

$$\frac{dT}{dt} + k^2a^2T = 0$$
$$\frac{d^2X}{dx^2} + k^2X = 0 \tag{6-7.5}$$

The two ordinary differential equation immediately yield the solutions

$$T(t) = Ae^{-k^2a^2t}$$
$$X(x) = C\cos kx + F\sin kx \tag{6-7.6}$$

Thus, the **particular solution**, (6-7.2) becomes

$$u(x, t) = X(x)T(t)$$

$$= e^{-k^2 a^2 t}\left[C\cos kx + F\sin kx\right]$$

(6-7.7)

Since the constants are arbitrary, we have retained two constants without changing the correcting of solution.

(b) Boundary and Initial conditions

Boundary conditions

The three constants C, F, and k are determined from the three boundary and initial conditions given by equations (6-7.1b and 1c), as follows.

Equation (6-7.1b), at x = 0 and x = l, gives upon substituting in equation (6-7.7)

$$u(0, t) = e^{-k^2 a^2 t}\left[C\cos kx\right] = 0$$

(6-7.8a)

Thus, C must vanish since the term containing functions in t cannot vanish at all times.

At x = l, we get

$$u(l, t) = e^{-k^2 a^2 t}\left[F\sin kl\right] = 0$$

(6-7.8b)

The roots of the sine term imply the following

$$kl = 0, \pm\pi, \pm n\pi$$

$$k = \frac{n\pi}{l}, \qquad n = 0, 1, 2, \ldots$$

(6-7.8c)

Thus, after satisfying the boundary conditions, we have

$$u(x, t) = e^{-\left(\frac{n\pi a}{l}\right)^2 t}\left[F\sin\frac{n\pi x}{l}\right]$$

(6-7.8d)

Initial conditions

Equation (6-7.1c), at t = 0 and t → ∞ and all x, gives upon substituting in equation (6-7.8d) and its derivative the following

$$\frac{\partial u(x, \infty)}{\partial t} = -\left(\frac{n\pi a}{l}\right)^2 e^{-\left(\frac{n\pi a}{l}\right)^2 \infty}\left[F\sin\frac{n\pi x}{l}\right] = 0, \qquad t \to \infty$$

(6-7.9a)

248

$$u(x,0) = \sum_{n=0}^{\infty} F_n \sin \frac{n\pi x}{l} = x, \qquad 0 \le x \le l \tag{6-7.9b}$$

Determining F_n by Euler's formulas:

Multiply both sides of equation (6-7.9) by $\sin \dfrac{m\pi x}{l}$, then integrate both sides with respect to x over the period 0 to $+l$, we get

$$\int_{x=0}^{l} \sum_{n=0}^{\infty} F_n \sin \frac{m\pi x}{l} \sin \frac{n\pi x}{l} dx = \int_{x=0}^{l} x \sin \frac{m\pi x}{l} dx \tag{6-7.10a}$$

We will now evaluate the following integral in the two cases: $i = j$ and $i \ne j$:

First, let us substitute the product of two sine-functions by the sum of two cosines, integrate, and substitute by the integration limits, as follows

$$\int_{x=0}^{l} F_n \sin \frac{m\pi x}{l} \sin \frac{n\pi x}{l} dx = \frac{1}{2} F_n \int_{x=0}^{l} \left(\cos \frac{m-n}{l}\pi x - \cos \frac{m+n}{l}\pi x \right) dx$$

$$= \frac{1}{2} F_n \left(\frac{\sin \dfrac{m-n}{l}\pi x}{\dfrac{m-n}{l}\pi} - \frac{\sin \dfrac{m+n}{l}\pi x}{\dfrac{m+n}{l}\pi} \right)_{x=0}^{x=l} \tag{6-7.10b}$$

Case 1: $n = m$

In equation (6-7.10b), it is easy to prove the following two relations

$$\lim_{(m-n)\to 0} \left(\frac{\sin(m-n)\pi}{\dfrac{m-n}{l}\pi} \right) = l$$

$$\lim_{(m+n)\to 0,1,2,3,} \left(\frac{\sin(m+n)\pi}{\dfrac{m+n}{l}\pi} \right) = 0 \tag{6-7.10c}$$

Thus, from (6-7.10a) and (6-7.10b), we get the first Euler's formula for the constants F_n's, as follows.

$$\int_{x=0}^{l} F_n \sin \frac{m\pi x}{l} \sin \frac{n\pi x}{l} dx = \frac{l}{2} F_n \qquad (6\text{-}7.10d)$$

Therefore,

$$F_n = \frac{2}{l} \int_{x=0}^{l} x \sin \frac{m\pi x}{l} dx \qquad (6\text{-}7.10e)$$

Integrating by parts, we get

$$F_n = -\frac{2}{n\pi} \int_{x=0}^{l} x d\cos \frac{n\pi x}{l}$$

$$= -\frac{2}{n\pi} \left(\left(x\cos\frac{n\pi x}{l} \right)_{x=0}^{l} - \int_{x=0}^{l} \cos\frac{n\pi x}{l} dx \right)$$

$$= -\frac{2}{n\pi} \left(\left(x\cos\frac{n\pi x}{l} \right)_{x=0}^{l} - \frac{l}{n\pi}\left(\sin\frac{n\pi x}{l} \right)_{x=0}^{l} \right) \qquad (6\text{-}7.10f)$$

$$= -\frac{2l}{n\pi} \left(\cos n\pi - \frac{1}{n\pi}\sin n\pi \right)$$

$$= -\frac{2l}{n\pi} \left(\cos n\pi \right)$$

Since the cosine changes sign every half cycle, we could express F_n as follows

$$F_n = (-1)^n \frac{2l}{n\pi}, \qquad \text{where } n = 1, 2, .. \qquad (6\text{-}7.10f)$$

We have excluded **n = 0** from our solution by virtue of equations (6-7.7) and (6-7.4). Of course, we could assume k = 0 in equation (6-7.4) and pursue such particular solution, which should be added to solution obtained at the end of this example.

Case 2: $n \neq m$

In equation (6-7.10b), it is easy to prove the following two relations

$$\lim_{(m-n)\to 0} \left| \frac{\sin(m-n)\pi}{\dfrac{m-n}{l}\pi} \right| = 0$$

$$\lim_{(m+n)\to 0,1,2,3,} \left| \frac{\sin(m+n)\pi}{\dfrac{m+n}{l}\pi} \right| = 0$$

(6-7.10g)

General Solution

From equation (6-7.8d) and (6-7.10f), the general solution of (6-7.1a) is

$$F_n = (-1)^{n+1} \frac{2l}{n\pi}$$

(6-7.11)

$$u(x,t) = \frac{2l}{\pi} \left[\sum_{n=1}^{\infty} (-1)^{n+1} \frac{2l}{n\pi} \sin\frac{n\pi x}{l} e^{-\left(\frac{n\pi a}{l}\right)^2 t} \right]$$

(6-7.8d)

The solution for $n = 0$ is left for the reader to entertain.

Example 95: Laplace Equation

Solve the Laplace's equation

e.g.,
$$\frac{\partial^2 u}{\partial x^2} + \frac{\partial^2 u}{\partial y^2} = 0$$

(6-8.1a)

Given the boundary conditions:

$$u(x,0) = u(x,\pi) = 0, \qquad \text{for all x} \qquad (6-8.1b)$$
$$u(\infty, y) = 0 \qquad \text{as } x \to \infty \qquad (6-8.1c)$$

Solution

(a) Separation of variables

Consider the solution

$$u(x,y) = X(x)Y(y)$$

(6-8.2)

Differentiate the above solution with respect to each of the independent variables x and y and substitute in the initial equation (6-8.1) to get

$$X \frac{d^2Y}{dy^2} + Y \frac{d^2X}{dx^2} = 0 \qquad (6-8.3a)$$

Divide both sides by XY, we get

$$\frac{1}{Y} \frac{d^2Y}{dy^2} = -\frac{1}{X} \frac{d^2X}{dx^2} \qquad (6-8.3b)$$

Since each side of the above equation is independent of the other side in regard to its variable, we have

$$\frac{1}{Y} \frac{d^2Y}{dy^2} = -\frac{1}{X} \frac{d^2X}{dx^2} = k^2 \qquad (6-8.4)$$

Where k is an arbitrary constant of proportionality and the negative sign is suggested to ease solution without affecting the outcome.

We could then separate the two sides of the above equations into two independent equations as follows

$$\frac{d^2Y}{dy^2} - Yk^2 = 0$$

$$\frac{d^2X}{dx^2} + Xk^2 = 0 \qquad (6-8.5)$$

The characteristic equations of the above two o.d.e. are

$$m^2 - k^2 = 0$$
$$n^2 + k^2 = 0 \qquad (6-8.6)$$

With the roots

$$m = \pm k$$
$$n = \pm ik \qquad (6-8.7)$$

With the four roots of the two characteristic equations for X and Y, equation (6-8.2) becomes

$$u(x, y) = X(x)Y(y)$$
$$= \left(Ae^{-kx} + Be^{kx} \right)\left(C\cos ky + F\sin ky \right) \qquad (6-8.8)$$

(b) Boundary and Initial conditions

252

Boundary conditions

Three of the four constants A, B, C and F, (k is included in one of the four constants) are determined from the given three boundary and initial conditions, equations (6-8.1b and 1c), as follows.

Equation (6-8.1b), at y = 0 and y = π, gives upon substituting in equation (6-8.8)

$$u(x,0) = \left(Ae^{-kx} + Be^{kx}\right)(C) \tag{6-8.9a}$$

Thus, C must vanish since the term containing functions in x cannot vanish at all x.

At y = π, we get

$$u(x,\pi) = \left(Ae^{-kx} + Be^{kx}\right)(F\sin k\pi) = 0 \tag{6-8.9b}$$

The roots of the sine term imply the following

$$k = 0,1,2,...,n \tag{6-8.9d}$$

Thus, after satisfying the boundary conditions, we have

$$u(x,y) = \sum_{k=0}^{\infty} \left(Ae^{-kx} + Be^{kx}\right)(\sin ky) \tag{6-8.9e}$$

Initial conditions

Equation (6-7.1c), at x → ∞ and all y, gives upon substituting in equation (6-8.9e) the following

$$u(\infty,y) = \sum_{k=0}^{\infty} \left(Ae^{-k\infty} + Be^{k\infty}\right)(\sin ky) \tag{6-8.9f}$$

Therefore, B must vanish in order to exclude the infinite value of the positive exponent.

Thus, we get

$$u(x,y) = \sum_{k=0}^{\infty} \left(A_k e^{k-\infty} \sin ky\right) \tag{6-8.9g}$$

The remaining fourth constant A_k requires additional boundary conditions, which are not given in the example. Hence, the above equation constitutes the required general solution of Laplace's equation, (6-8.1)

Example 96: Wave Equation

Solve the p.d.e. Equation

e.g.,
$$\frac{\partial^2 u}{\partial x^2} + 12\frac{\partial u}{\partial x} = 9\frac{\partial^2 u}{\partial t^2} \qquad\qquad (6\text{-}9.1a)$$

Given the boundary conditions:

$$u(0, t) = \cos 6t, \qquad\qquad\qquad \text{for all } t \qquad\qquad (6\text{-}9.1b)$$
$$\frac{\partial u(0, t)}{\partial x} = 0 \qquad\qquad\qquad\qquad (6\text{-}9.1c)$$

Solution

(a) Separation of variables

Consider the solution

$$u(x, y) = X(x)T(t) \qquad\qquad (6\text{-}9.2)$$

Differentiate the above solution with respect to each of the independent variables x and y and substitute in the initial equation (6-9.1) to get

$$T\frac{d^2X}{dx^2} + 12T\frac{dX}{dx} = 9X\frac{d^2T}{dt^2} \qquad\qquad (6\text{-}9.3a)$$

Divide both sides by 9XT, we get

$$\frac{1}{9X}\frac{d^2X}{dx^2} + 12\frac{1}{9X}\frac{dX}{dx} = \frac{1}{T}\frac{d^2T}{dt^2} \qquad\qquad (6\text{-}9.3b)$$

Since each side of the above equation is independent of the other side in regard to its variable, we have

$$\frac{1}{9X}\frac{d^2X}{dx^2} + 12\frac{1}{9X}\frac{dX}{dx} = \frac{1}{T}\frac{d^2T}{dt^2} = -k^2 \qquad\qquad (6\text{-}9.4)$$

Where k is an arbitrary constant of proportionality and the negative sign is suggested to ease solution without affecting the outcome.

We could then separate the two sides of the above equations into two independent equations as follows

$$\frac{1}{9X}\frac{d^2X}{dx^2} + 12\frac{1}{9X}\frac{dX}{dx} = -k^2$$

$$\frac{1}{T}\frac{dT}{dt} = -k^2$$

(6-9.5a)

Or.

$$\frac{d^2X}{dx^2} + 12\frac{dX}{dx} + 9k^2X = 0$$

$$\frac{d^2T}{dt^2} + k^2T = 0$$

(6-9.5b)

The characteristic equations of the above two o.d.e. are

$$m^2 + 12m + 9k^2 = 0$$

$$n^2 + k^2 = 0$$

(6-9.6)

With the roots

$$m = \frac{-12 \pm \sqrt{144 - 36k^2}}{2}$$

$$n = \pm ik$$

(6-9.7a)

First, we should organize the terms in the m roots as follows

$$m = -6 \pm \sqrt{36 - 9k^2}$$

$$= -6 \pm i\sqrt{9k^2 - 36}$$

$$n = \pm ik$$

(6-9.7b)

Those four roots of the two characteristic equations for X and T yield the solution, equation (6-9.2)

$$u(x, y) = X(x)T(t)$$

$$= e^{-6x}\left[A\cos\sqrt{9k^2 - 36}x + B\sin\sqrt{9k^2 - 36}x\right]\left(C\cos kt + F\sin kt\right)$$

(6-9.8)

(b) Boundary and Initial conditions

255

Boundary conditions

Two of the four constants A, B, C and F, (k is included in one of the four constants) are determined from the given two boundary and initial conditions, equations (6-9.1b and 1c), as follows.

Equation (6-9.1b), **at x = 0,** gives upon substituting in equation (6-9.8) gives

$$u(0,t) = A(C\cos kt + F\sin kt) = \cos 6t \qquad (6\text{-}9.9a)$$

Determining k by Euler's formulas:

Multiply both sides of equation (6-9.9a) by $\cos nt$, then integrate both sides with respect to t over the period 0 to $+l$, we get

Noting the following trigonometric relationships:

$$
\begin{aligned}
\cos(\alpha + \beta) &= \cos\alpha\cos\beta - \sin\alpha\sin\beta \\
\cos(\alpha - \beta) &= \cos\alpha\cos\beta + \sin\alpha\sin\beta \\
\sin(\alpha + \beta) &= \sin\alpha\cos\beta + \cos\alpha\sin\beta \\
\sin(\alpha - \beta) &= \sin\alpha\cos\beta - \cos\alpha\sin\beta
\end{aligned}
\qquad (6\text{-}9.10a)
$$

$$
\begin{aligned}
\int_{x=0}^{l} A(C\cos kt + F\sin kt)\cos nt\, dt &= \int_{x=0}^{l} A(C\cos kt\cos nt + F\sin kt\cos nt)dt \\
&= \int_{x=0}^{l} A(C\cos kt\cos nt + F\sin kt\cos nt)dt \\
&= \int_{x=0}^{l} A\left(C\left[\dfrac{\cos(k+n)t + \cos(k-n)t}{2}\right] + F\left[\dfrac{\sin(k+n)t + \sin(k-n)t}{2}\right] \right)dt
\end{aligned}
\qquad (6\text{-}9.10b)
$$

Integrating, we get

$$
\begin{aligned}
&= \int_{x=0}^{l} A\left(C\left[\dfrac{\cos(k+n)t + \cos(k-n)t}{2}\right] + F\left[\dfrac{\sin(k+n)t + \sin(k-n)t}{2}\right] \right)dt \\
&= \frac{1}{2}\int_{x=0}^{l} A\left(C\left[\dfrac{\sin(k+n)t}{(k+n)} + \dfrac{\sin(k-n)t}{(k-n)}\right] - F\left[\dfrac{\cos(k+n)t}{(k+n)} + \dfrac{\cos(k-n)t}{(k-n)}\right] \right)dt
\end{aligned}
$$

$$(6\text{-}9.10c)$$

As we have proved in equation (6-7.10), we can easily prove that, in equation (6-9.9a), the following holds true

$$k = 6$$
$$F = 0 \qquad\qquad\qquad (6-9.10d)$$
$$AC = 1$$

Equation (6-9.1c), at $x = 0$, gives upon substituting in the derivative of equation (6-9.8) gives

$$\frac{\partial u(0,t)}{\partial x} = \left[-6A + B\sqrt{9k^2 - 36}\right](C\cos kt + F\sin kt) = 0 \qquad\qquad (6-9.10e)$$

Therefore, since the term including t does not vanish, we get

$$A = \frac{B\sqrt{9k^2 - 36}}{6} \qquad\qquad\qquad (6-9.10f)$$

Substituting by k = 6, from (6-9.10d), we get

$$B = \frac{A}{2\sqrt{2}} \qquad\qquad\qquad (6-9.10g)$$

Thus, the **general solution** is

$$u(x,y) = X(x)T(t)$$
$$= e^{-6x}\left[\cos 12\sqrt{2}x + \frac{1}{2\sqrt{2}}\sin 12\sqrt{2}x\right]\cos 6t \qquad\qquad (6-9.11)$$

6.4. Exercises on p.d.e.

6-1. Solve the Laplace's equation

$$\frac{\partial^2 u}{\partial x^2} + \frac{\partial^2 u}{\partial y^2} = 0$$

Given the boundary conditions:

$$u(0, y) = u(\pi, y) = 0, \qquad\qquad \text{for all y}$$
$$u(x, \infty) = 0 \qquad\qquad \text{as y} \to \infty$$

6-2. Find a solution for the p.d.e.

$$\frac{\partial^2 u}{\partial x^2} = \frac{\partial u}{\partial t}$$

Given the boundary conditions:

$$u(0, t) = u(\pi, t) = 0, \qquad\qquad \text{for all t}$$
$$u(x, 0) = 2\sin x - 3\sin 3x$$

6-3. Find a solution for the p.d.e.

$$\frac{\partial^2 u}{\partial t^2} = a^2 \frac{\partial^2 u}{\partial x^2}$$

Given the boundary conditions:
$$u(0, t) = u(l, t) = 0, \qquad\qquad \text{for all t}$$

And initial conditions

$$u(x, 0) = 2\sin \frac{3\pi}{l} x$$
$$\frac{\partial u(x, 0)}{\partial t} = 0 \qquad\qquad \text{for all } 0 \le x \le l$$

Find the deflection function at any time t and distance x.

6-4. Solve the Laplace's equation

$$\frac{\partial^2 u}{\partial x^2} + \frac{\partial^2 u}{\partial y^2} = 0$$

In the semi-infinite strip $0 \le x \le a$ and $0 \le y \le \infty$, given the boundary conditions:

$u(0, y) = u(a, y) = 0$,	for all y
$u(x, \infty) = 0$	as $y \to \infty$
$u(x,0) = A\left(1 - \dfrac{x}{a}\right)$	as for all values of x in the strip.

6-5. Solve the equation

$$\frac{\partial^2 V}{\partial x^2} + \frac{\partial^2 V}{\partial y^2} + 7V = 0$$

Given the boundary conditions:

a)   V is periodic in x
b)   V = 0 when x = 0 for all values of y

c)   ☒

6-6. The temperature u(x,t) at a point on a conducting rod, distant x from one of its ends at time t satisfies the p.d.e.

$$\frac{\partial u}{\partial t} = a^2 \frac{\partial^2 u}{\partial x^2}$$

Where a is constant.

Find the temperature distribution that satisfies the conditions

a) u is finite at all times.
b) u = 0 at x = 0 and for t ≥ 0
c) $\dfrac{\partial u}{\partial x} = 0$ at x = l and for t ≥ 0
d) u = x when t = 0 and for all x in the interval $0 \le x \le l$
at x = 0 and for t ≥ 0

6-7. Solve the p.d.e. equation

e.g., $$\frac{\partial^2 u}{\partial t^2} + 2k \frac{\partial u}{\partial t} = c^2 \frac{\partial^2 u}{\partial x^2}$$

Given the boundary conditions:

$$u(x,0) = \cos \frac{3kx}{c}$$

$$\frac{\partial u(x,0)}{\partial t} = 0$$

CHAPTER 7

APPLICATIONS ON O.D.E & P.D.E

7-1. Applications on O.D.E.

7-1.1. Differential equation of electric current and voltage of RCL circuits

Electric circuits containing resistance R, capacitance C, inductance L performs the following tasks:

a) Stores electric charge Q (coulombs)
b) Conduct current I (ampere)
c) Drops voltage V_R (volt) on the resistance R
d) Drops voltage V_C (volt) on the capacitance C
e) Drops voltage V_L (volt) on the inductance L

The **governing laws** are

Current: $$I(ampere) = \frac{dQ(coulomb)}{dt(second)}$$ (7-1.1)

Voltage Drop: $$V_L(volt) = -L(henries)\frac{dI(ampere)}{dt(second)}$$ (7-1.2)

Voltage Drop: ☒   (7-1.3)

Voltage Drop: ☒   (7-1.4)

Summary

Current   Induction loss   Resistive loss   Capacitive loss

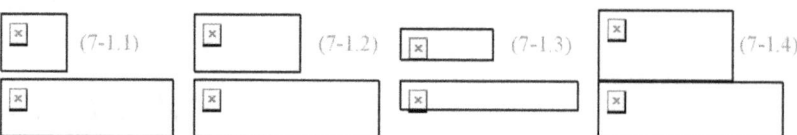

(7-1.1)   (7-1.2)   (7-1.3)   (7-1.4)

Energy balance in a circuit

The energizing source E, e.m.f. (electromotive force) is given by

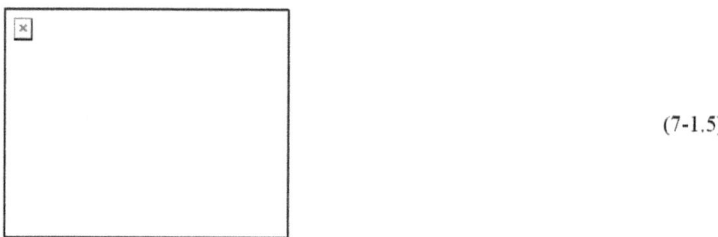

(7-1.5)

The current form of (7-1.5) is obtained by differentiating it w.r.t. time and substituting from (7-1.1), to get

(7-1.6)

Summary of voltage equilibrium in RCL electric circuit

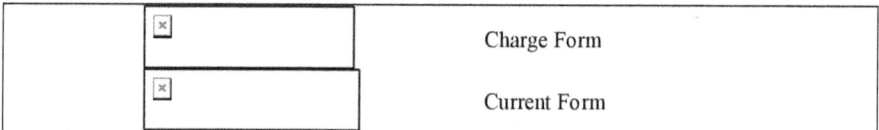

	Charge Form
	Current Form

7-1.1a. Kirchoff's voltage law

(a) Junction point zero net current

Any junction of wires in an electric circuit, the net sum of currents (**positive towards the joint, negative outwards**) is zero.

(b) Closed loop zero net sum e.m.f.

Any closed loop in an electric circuit, the net sum of voltages (**positive clockwise, negative counterclockwise**) is zero.

7-1.2. Differential equation of mechanical vibration

State of material motion

(i) Free oscillation or homogeneous o.d.e.

Example 97

The equation of motion of **mass** m of material object, translated **distance** y (deflection) from equilibrium position. (y is taken positive downwards, negative upwards). Figure 7-1.

$$(7\text{-}2.1)$$

Force applied to mass (restoring force) is imparted by spring elastic pull, is proportional to **deflection** y. (k is proportionality constant).

$$(7\text{-}2.2)$$

Motion is exposed to **resistive force** proportional to the speed of motion (r is proportionality constant or shock absorber).

$$(7\text{-}2.3)$$

Newton's second law of motion

$$(7\text{-}2.4)$$

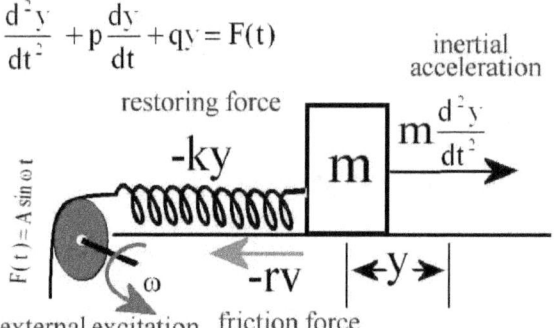

$$\frac{d^2y}{dt^2} + p\frac{dy}{dt} + qy = F(t)$$

inertial acceleration

restoring force

$$-ky \qquad m \qquad m\frac{d^2y}{dt^2}$$

$F(t) = A\sin\omega t$

ω

$-rv$ $|\leftarrow y \rightarrow|$

external excitation   friction force

Figure 7-1. State of motion of mass m under resistive forces -rv and restoring force -ky excited by sinusoidal external force with frequency ω.

The equation of motion (7-2.4) is written in the general form as follows:

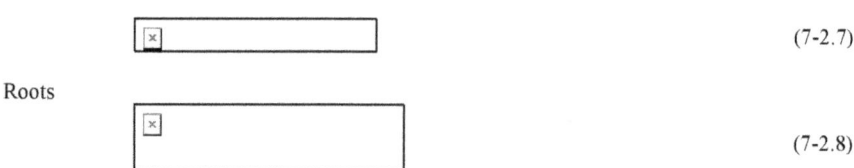

(7-2.5)

i.e.

(7-2.6)

The **homogenous** o.d.e. represents **free oscillations** in which restoring and resistive forces account for the acceleration of the material mass.

Characteristic equation

(7-2.7)

Roots

(7-2.8)

Possible states of motion, Figure 7-2

Case # 1:	Resistance greater than restoring force

The roots of the characteristic equation are real, distinct and negative.

The deflection distance and solution of (7-2.6) and the electric current and solution of (7-1.6) are

264

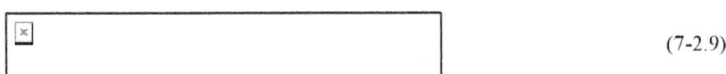

(7-2.9)

Since the two roots m_1 and m_2, are negative, both deflection and electric current decay as time grows due to the greater resistance.

Case # 2: 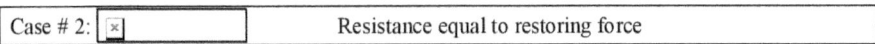 Resistance equal to restoring force

The roots of the characteristic equation are real, equal and negative.

The deflection distance and solution of (7-2.6) and the electric current and solution of (7-1.6) are

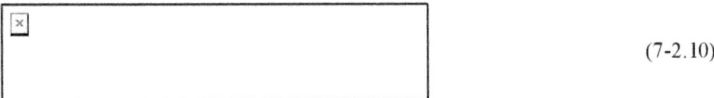

(7-2.10)

Deflection and electric do not decay as fast as in case #1 since the linear growth competes with the exponential decay.

Case # 3: 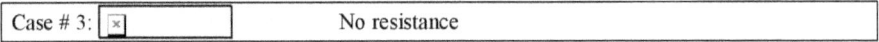 No resistance

The roots of the characteristic equation are imaginary and distinct.

The deflection distance and solution of (7-2.6) and the electric current and solution of (7-1.6) are

(7-2.11)

This can be written in the more representative form of wave propagation with **phase shift** as follows.

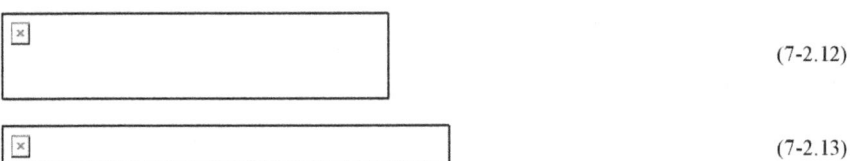

(7-2.12)

(7-2.13)

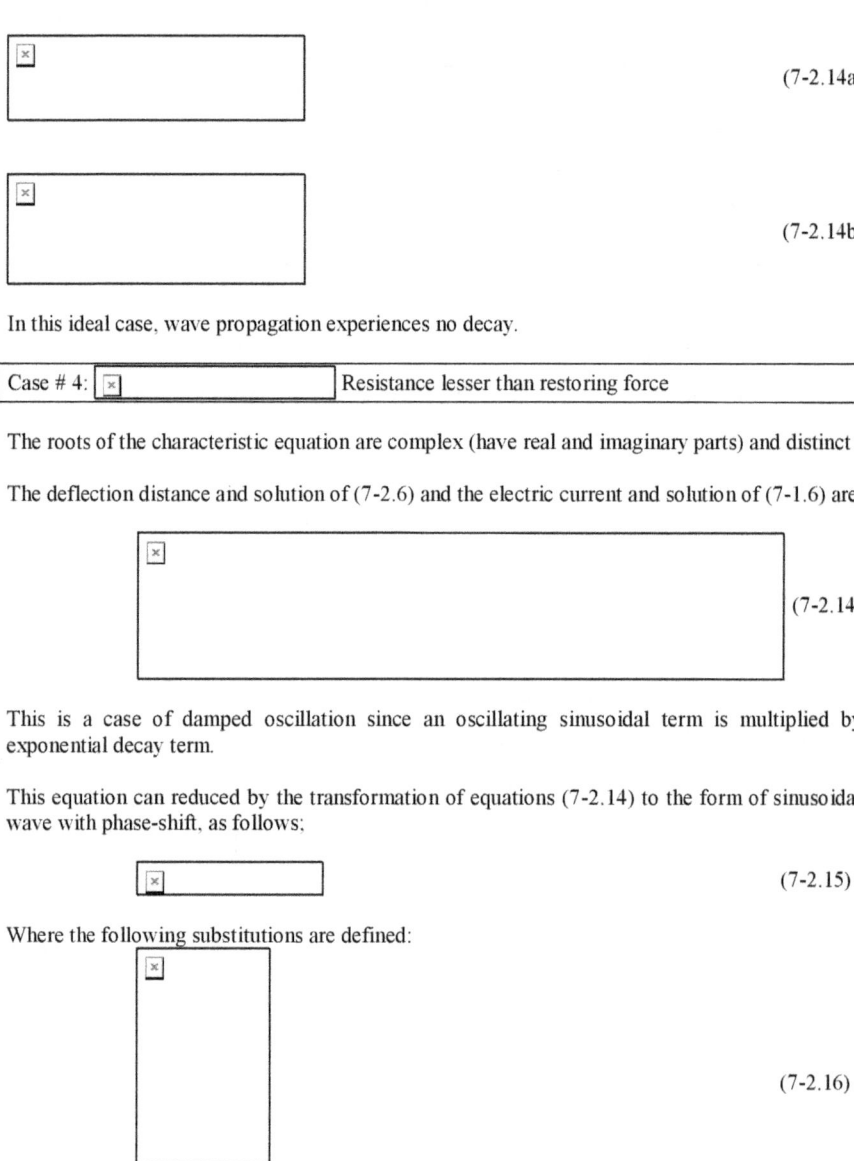

$$(7\text{-}2.14a)$$

$$(7\text{-}2.14b)$$

In this ideal case, wave propagation experiences no decay.

Case # 4:	Resistance lesser than restoring force

The roots of the characteristic equation are complex (have real and imaginary parts) and distinct

The deflection distance and solution of (7-2.6) and the electric current and solution of (7-1.6) are

$$(7\text{-}2.14)$$

This is a case of damped oscillation since an oscillating sinusoidal term is multiplied by exponential decay term.

This equation can reduced by the transformation of equations (7-2.14) to the form of sinusoidal wave with phase-shift, as follows;

$$(7\text{-}2.15)$$

Where the following substitutions are defined:

$$(7\text{-}2.16)$$

266

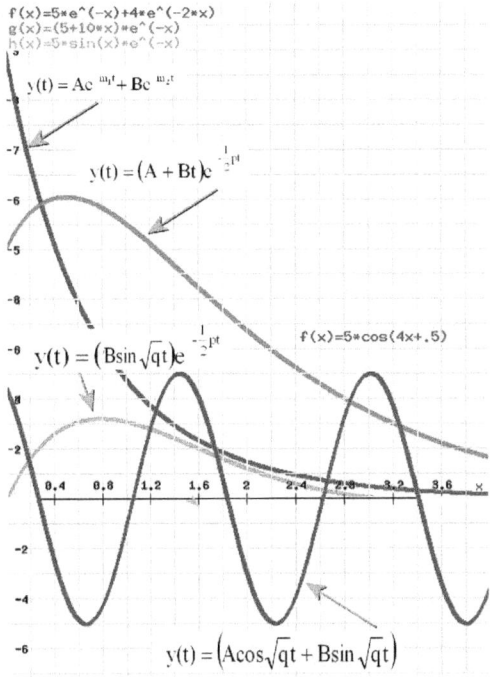

$f(x)=5*e^{(-x)}+4*e^{(-2*x)}$
$g(x)=(5+10*x)*e^{(-x)}$
$h(x)=5*sin(x)*e^{(-x)}$

$y(t) = Ae^{m_1 t} + Be^{m_2 t}$

$y(t) = (A + Bt)e^{-\frac{1}{2}t^4}$

$y(t) = (B\sin\sqrt{q}t)e^{-\frac{1}{2}pt}$

$f(x)=5*\cos(4x+.5)$

$y(t) = (A\cos\sqrt{q}t + B\sin\sqrt{q}t)$

Figure 7-2. Four cases of motion depending on the roots of the characteristic equation (7-2.6).

(ii) Forced oscillation or nonhomogeneous o.d.e.

Example 98

This occurs, for example, when the restoring force **depends on time t**, not only distance y (such as spring attached to roller moving on uneven surface). In this case, the restoring force takes the form.

$$\boxed{\times } \tag{7-3.1}$$

The **resistive force** changes to

267

$$\text{resistive force} = -r\frac{dy}{dt}[y + f(t)] \qquad (7\text{-}3.2)$$

Newton's second law of motion gives

$$\text{resistive force} = -r\frac{d[y + f(t)]}{dt} \qquad (7\text{-}3.3)$$

The equation of motion (7-2.4) is written in the general form as follows:

$$\frac{d^2y}{dt^2} + \frac{r}{m}\frac{d(y + f(t))}{dt} + \frac{k(y + f(t))}{m} = 0$$
$$\frac{d^2y}{dt^2} + \frac{r}{m}\frac{dy}{dt} + \frac{ky}{m} = -\frac{r}{m}\frac{df(t)}{dt} - \frac{kf(t)}{m} \qquad (7\text{-}3.4)$$

i.e.,
$$\frac{d^2y}{dt^2} + p\frac{dy}{dt} + qy = F(t) \qquad (7\text{-}3.5)$$

Where,

$$F(t) = -\frac{r}{m}\frac{df(t)}{dt} - \frac{kf(t)}{m} \qquad (7\text{-}3.6)$$

The **nonhomogenous** o.d.e. represents **forced oscillations** in which restoring and resistive forces are modified by the function f(t), in equation (7-3.1).

Periodic external excitation

Consider the excitation function in equation (7-3.5) of the form

$$F(t) = A \sin \omega t \qquad (7\text{-}3.7)$$

First, consider [×]

The o.d.e. of motion with forced oscillation becomes

$$\frac{d^2y}{dt^2} + p\frac{dy}{dt} + qy = A\sin\omega t \qquad (7\text{-}3.8)$$

The **complementary function** of the homogeneous equation of (7-3.8) is obtained in (7-2.15)

$$\text{C.F.} = He^{\alpha}\sin(\beta t + \varphi) \tag{7-3.9a}$$

Where α is given by (7-2.16)

$$\alpha = -\frac{1}{2}p \tag{7-3.9b}$$

$$\beta = \sqrt{q - \frac{1}{4}p^2} \tag{7-3.9c}$$

The **particular solution** of the **nonhomogeneous** equation of (7-3.8) is obtained by first putting in terms of D-Operator as follows:

$$\left(D^2 + pD + q\right)y = A\sin\omega t$$

$$y = \frac{1}{\left(D^2 + pD + q\right)}A\sin\omega t \tag{7-3.10a}$$

As we have done before, we can use **Euler's substitution** for the sine function as follows

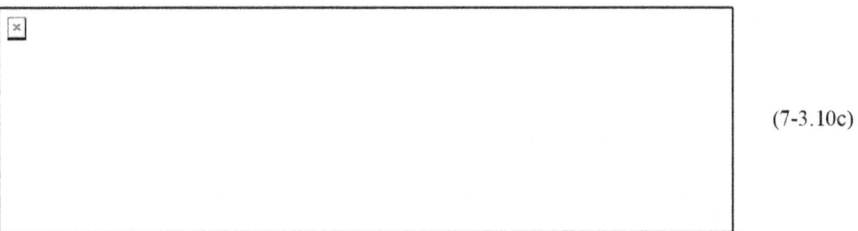

$$\tag{7-3.10b}$$

We have used equation (3-49.6) which deals with **Inverse-D-Operator** of exponential function.

After performing the arrangement, the Imaginary part comprises our sought solution.

This could be farther arranged as follows

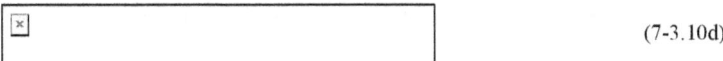

$$\tag{7-3.10c}$$

i.e.,

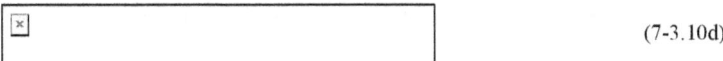

$$\tag{7-3.10d}$$

We will make substitutions similar to (7-2.14) in order to arrange the above equation in the phase-shift form as follows

(7-3.10e)

Where,

(7-3.10f)

(7-3.10g)

General solution is sum of complementary function (7-3.9) and particular solution equation (7-3.10g):

$$G.S. = C.F. + P.S.$$
$$= H e^{\alpha} \sin(\beta t + \varphi) + G \sin(\omega t + \psi)$$

(7-3.10h)

(ii.a) Analysis of forced oscillation

The solution of forced oscillation, equation (7-3.10h) can be described as follows

$$y(t) = \underset{\text{transient and decay}}{H e^{\alpha} \sin(\beta t + \varphi)} + \underset{\text{steady state}}{G \sin(\omega t + \psi)}$$

(7-3.11a)

1. The C.F. comprises the decaying sinusoidal wave, which diminishes as time t grows. Thus, the C.F., comprises the transient phase.

2. The P.S. comprises the steady-state wave without decay.

3. The magnitude G, equation (7-3.10f), of the forced P.S. can be written in the form

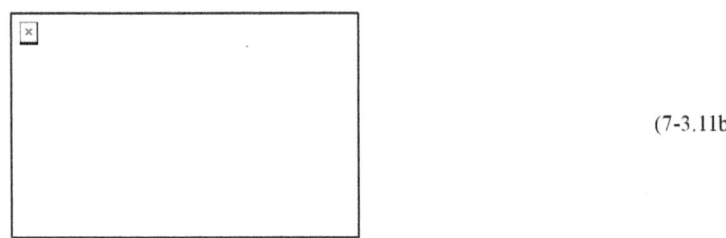

(7-3.11b)

Since p comprises **resistive forces**, q **restoring forces**, and ω the frequency of **externally forced** oscillation, we note that if **p = 0**, G becomes maximum.

(7-3.11c)

4. **Resonance** occurs when the $q - \omega^2$

(iii) Comparison between electric and mechanical motion

	Electrical		Mechanical		

resistance	R		R		friction
induction	L		M		mass
Induction-capacitance	LC		m		mass
current	I		y		deflection
e.m.f.	(dE/dt)/L		F(t)		force

7-1.3. Buckling of elastic rod (Euler's problem)

Example 99

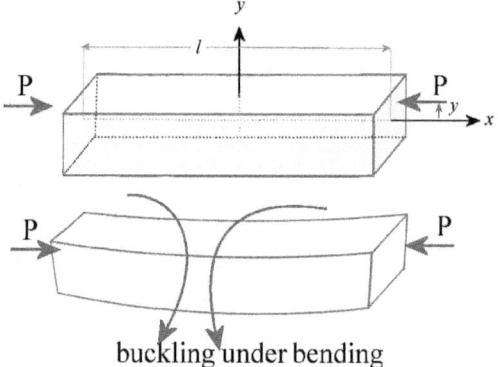
buckling under bending

Figure 7-3. Buckling of rod le length l, moment of inertia J, and Young's modulus E under axial forces acting along the x axis.

Euler's bending problem comprises of buckling of a bar of length l due to the action of compression forces acting on the two ends of the bar. Figure 7-3.

Since the bar has thickness (say, h) along the y-axis (perpendicular to the direction of forces), the axial forces exert **bending couple** given by

$$EJ \frac{d^2y}{dx^2} = -Py \qquad (7\text{-}4.1)$$

Here, the Young's modulus E is gotten from Hooke's law of elasticity

$$\sigma = E\varepsilon \qquad (7\text{-}4.2)$$

where,

σ denotes stress, defined by force per unit area of surface perpendicular to direction of force
ε denotes strain, defined by elongation per unit length along the direction of force.
J denotes the moment of inertia of the rod about the axis of rotation
The second derivative of y represents the curvature of the rod.

Equation (7-4.1) is written in our convenient notation as

$$\frac{d^2y}{dt^2} + \frac{P}{EJ} y = 0 \qquad (7\text{-}4.3a)$$

The characteristic equation is

$$m^2 + \frac{P}{EJ} = 0 \qquad (7\text{-}4.3b)$$

Thus, since the two roots are imaginary, the general solution becomes

$$y(t) = A\sin\left(\sqrt{\frac{P}{EJ}}x\right) + B\cos\left(\sqrt{\frac{P}{EJ}}x\right) \qquad (7\text{-}4.4)$$

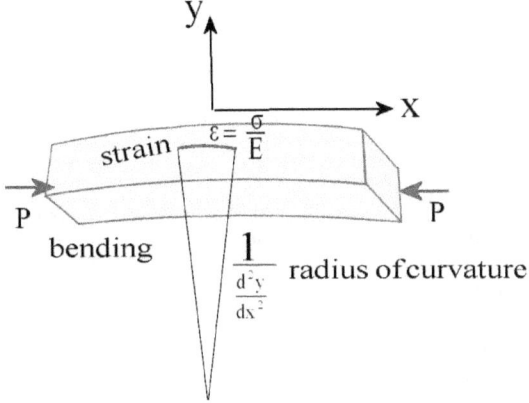

Figure 7-4. Curving of rod

Boundary conditions

Assuming that the two ends are fixed, then

$$y(0) = y(l) = 0 \qquad (7\text{-}4.5a)$$

Substituting by the boundary conditions in (7-4.4), we get

$$y(0) = B = 0$$
$$y(l) = A\sin\left(\sqrt{\frac{P}{EJ}}l\right) = 0 \qquad (7\text{-}4.5b)$$

The roots of the sine function give

$$\sqrt{\frac{P}{EJ}}\,l = n\pi, \qquad n = 0,1,2,...\infty \qquad\qquad (7\text{-}4.5c)$$

i.e.,

$$P_n = EJ\left(\frac{n\pi}{l}\right)^2 \qquad\qquad (7\text{-}4.5d)$$

The general solution is

$$y(x) = A\sin\left(\frac{n\pi x}{l}\right) \qquad\qquad (7\text{-}4.5e)$$

7-1.4. Whirling of elastic rod

Example 100

An elastic rod with length density w (grams per unit length), Young's modulus E is spun at angular speed ω and moment inertia J, and deflected distance y, Figure 7-5.

The **centrifugal force** on the length dx of mass w dx at speed ω is

$$\text{centrifugal force} = \text{mass} * \frac{(\text{linear velocity})^2}{\text{radius of orbit}}$$

$$= \left(\frac{wdx}{g}\right)\left(\frac{(y\omega)^2}{y}\right) \qquad\qquad (7\text{-}5.1)$$

$$dF = \frac{w}{g}y\omega^2 dx$$

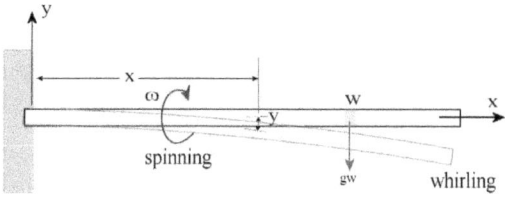

Figure 7-5. Whirling of spun rod

274

The **stress** along the y-direction is calculated by dividing the centrifugal force by the length dx.

$$\sigma_y = \frac{dF}{dx}$$
$$= \frac{w}{g} y\omega^2$$

(7-5.2)

Equation (7-4.1) gives the **bending moment** (M = Fy, force times deflection)

$$EJ\frac{d^2y}{dx^2} = \int Fxdx$$

(7-5.3)

$$= M$$

The bending moment is created by the centrifugal force bending the rod around its fixed end at distance x from fixation.

Thus, the **stress** σ_y is thus given by

$$\sigma_y(x) = \frac{d^2M}{dx^2}$$

(7-5.4a)

The double derivative of **moment of bending** with respect to x is explained as follows.
The first derivative of M gives the **differential bending moment** on the infinitesimal element dx.
The second derivative of M gives the **differential of force per unit length**, or stress.

Thus, from equation (7-5.2) and (7-5.3), we get

$$\sigma_y(x) = \frac{d^2}{dx^2}\left(EJ\frac{d^2y}{dx^2}\right)$$
$$= \frac{w}{g} y\omega^2$$

(7-5.4b)

This can be written in our convenient notation as follows

$$EJ\frac{d^4y}{dx^4} - \frac{w}{g}\omega^2 y = 0$$

(7-5.4c)

i.e.,
$$\left(D^4 - q^4\right)y = 0$$

(7-5.4d)

Where,

$$q = \left(\frac{w}{gEJ} \omega^2 \right)^{\frac{1}{4}}$$ 　　　　　　　(7-5.4e)

Equation (7-5.4d) is the governing equation of spinning motion of an elastic rod. Its **characteristic equation** is

$$(m - q)(m + q)(m^2 + q^2) = 0$$ 　　　　　　(7-5.4f)

Thus, we have two real distinct roots and two imaginary roots, giving the equation of deflection of the rod from initial position of equilibrium as follows

$$y(x) = A\cos(qx) + B\sin(qx) + C\sinh(qx) + F\cosh(qx)$$ 　　　(7-5.5)

Loading boundary conditions

The four constants A, B, C, and F are determined from four boundary conditions as follows.

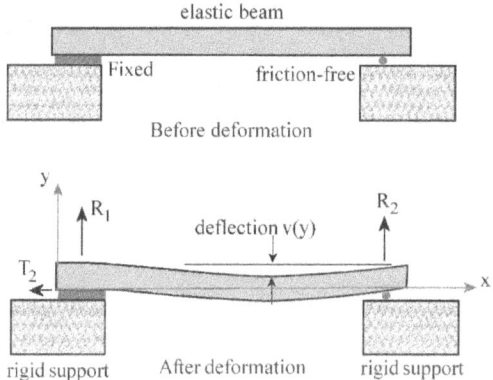

Figure 7-6. Loading condition at ends of a spinning bent bar

Case # 1: Rod supported on both ends by two short bearings

Vanishing **deflection** and **bending moment** at both ends.

$$y(0) = y(l) = 0$$

$$\left.\frac{d^2y}{dx^2}\right|_{x=0,l} = 0 \qquad\qquad (7\text{-}6.1)$$

Substituting by the boundary conditions in the deflection equation (7-5.5), we get

$$y(0) = A + F = 0$$
$$y(l) = A\cos(ql) + B\sin(ql) + C\sinh(ql) + F\cosh(ql) = 0$$
$$y''(0) = q^2[-A+F] = 0 \qquad\qquad (7\text{-}6.2a)$$
$$y''(l) = q^2[-A\cos(ql) - B\sin(ql) + C\sinh(ql) + F\cosh(ql)] = 0$$

Thus,

$$F = A = 0 \qquad\qquad (7\text{-}6.2b)$$

$$B\sin(ql) + C\sinh(ql) = 0$$
$$-B\sin(ql) + C\sinh(ql) = 0 \qquad\qquad (7\text{-}6.2c)$$

Therefore, if sin(ql) or sinh(ql) **do not vanish** by the proper choice of q and l, then

$$B = C = 0 \qquad\qquad (7\text{-}6.2d)$$

If sin(ql) or sinh(ql) **vanish** by the proper choice of q and l, then

$$\sin(ql) = 0$$
$$ql = n\pi \qquad\qquad (7\text{-}6.2e)$$

Substituting by the value of q_n in equation (7-5.4e), we get

$$\omega_n = \sqrt{\frac{gEJ}{W}}\left(\frac{n\pi}{l}\right)^2 \quad , n = 1,\, 2,\, 3,\, \dots \qquad\qquad (7\text{-}6.2f)$$

This is the **critical spinning speed** at which the bar passes from straight (zero deflection) to whirling motion when both ends are supported on **short bearings**.

Case # 2: Rod supported on both ends by two long bearings

Vanishing **deflection** and **tangent inclination or rotation** at both ends.

$$y(0) = y(l) = 0$$

$$\left.\frac{dy}{dx}\right|_{x=0,l} = 0 \qquad\qquad (7\text{-}7.1)$$

Substituting by the boundary conditions in the deflection equation (7-5.5), we get

$$y(0) = A + F = 0$$
$$y(l) = A\cos(ql) + B\sin(ql) + C\sinh(ql) + F\cosh(ql) = 0$$
$$y'(0) = q[B + C] = 0 \tag{7-7.2a}$$
$$y'(l) = q[-A\sin(ql) + B\cos(ql) + C\cosh(ql) + F\sinh(ql)] = 0$$

Thus,

$$F = -A \tag{7-7.2b}$$
$$B = -C \tag{7-7.2c}$$

$$A(\cos(ql) - \cosh(ql)) = B(-\sin(ql) + \sinh(ql))$$
$$A(-\sin(ql) - \sinh(ql)) = B(-\cos(ql) + \cosh(ql)) \tag{7-7.2d}$$

Dividing the two equations by members, we get

$$(\cos(ql) - \cosh(ql))(-\cos(ql) + \cosh(ql)) = (-\sin(ql) - \sinh(ql))(-\sin(ql) + \sinh(ql)) \tag{7-7.2e}$$

i.e.,

$$(\sinh^2(ql) - \sin^2(ql)) - (\cosh(ql) - \cos(ql))^2 = 0$$
$$(\sinh^2(ql) - \sin^2(ql)) - (\cosh^2(ql) - 2\cosh(ql)\cos(ql) + \cos^2(ql)) = 0 \tag{7-7.2f}$$
$$-2 + 2\cosh(ql)\cos(ql) = 0$$

Or

$$\cosh(ql)\cos(ql) = 1 \tag{7-7.3}$$

The roots of this equation could be shown graphically to approximate

$$ql = \left(n + \frac{1}{2}\right)\pi \tag{7-7.4}$$

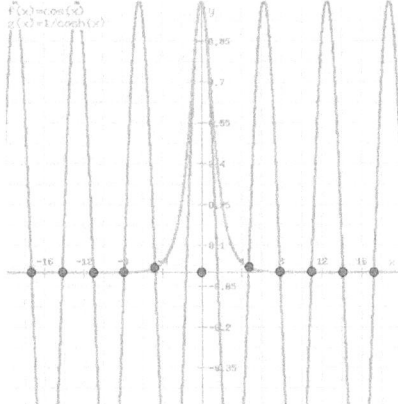

Figure 7-7. Roots of the transcendental equation (cos x cosh x =1).

Substituting by the value of q_n from (7-7.4) in equation (7-5.4e), we get

$$\omega_n = \sqrt{\frac{gEJ}{W}}\left(\left(n+\frac{1}{2}\right)\frac{\pi}{l}\right)^2 , n = 1, 2, \dots \qquad (7\text{-}7.5)$$

This is the **critical spinning speed** at which the bar passes from straight (zero deflection) to whirling motion when **both ends** are prevented from bending and deflecting on **long bearings**.

Case # 3: Rod supported on a short bearing at one end and a long bearing at the other end

Vanishing **deflection** at both ends, vanishing **tangent inclination** at one end and vanishing **moment of bending** at the other end.

$$y(0) = y(l) = 0$$
$$\left.\frac{dy}{dx}\right|_{x=0} = 0 \qquad\qquad (7\text{-}8.1)$$
$$\left.\frac{d^2y}{dx^2}\right|_{x=l} = 0$$

Substituting by the boundary conditions in the deflection equation (7-5.5), we get

279

$$y(0) = A + F = 0$$
$$y(l) = A\cos(ql) + B\sin(ql) + C\sinh(ql) + F\cosh(ql) = 0$$
$$y'(0) = q[B + C] = 0 \qquad\qquad (7\text{-}8.2a)$$
$$y''(l) = q^2[-A\cos(ql) - B\sin(ql) + C\sinh(ql) + F\cosh(ql)] = 0$$

Thus,

$$F = -A \qquad\qquad (7\text{-}8.2b)$$
$$B = -C \qquad\qquad (7\text{-}8.2c)$$

Like in the previous example, substituting by those in the remaining two equations, and dividing the two equations by members, we get

$$(\cos(ql) - \cosh(ql))(\sin(ql) + \sinh(ql)) = (\sin(ql) - \sinh(ql))(\cos(ql) + \cosh(ql))$$
$$(7\text{-}8.2d)$$

i.e.,

$$\boxed{\tanh(ql) = \tan(ql)} \qquad\qquad (7\text{-}8.2e)$$

The roots of this equation could be shown graphically to approximate

$$ql = \left(n + \frac{1}{4}\right)\pi \qquad\qquad (7\text{-}8.4)$$

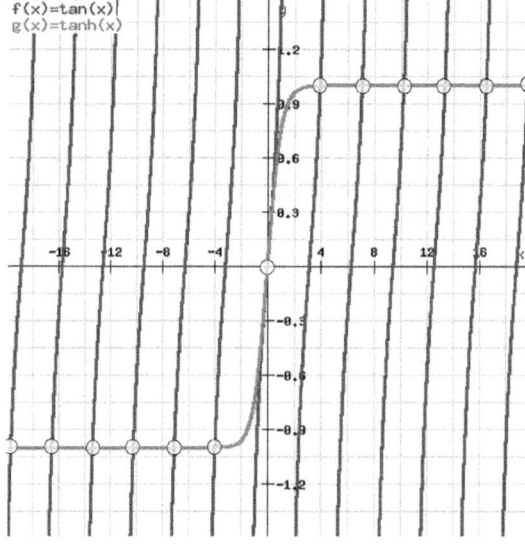

f(x)=tan(x)
g(x)=tanh(x)

Figure 7-8. Roots of the transcendental equation (tan x =tanh x).

Substituting by the value of q_n from (7-8.4) in equation (7-5.4e), we get

$$\omega_n = \sqrt{\frac{gEJ}{W}}\left(\left(n+\frac{1}{4}\right)\frac{\pi}{l}\right)^2 \,,\, (n = 1,\, 2,\, ...) \tag{7-8.5}$$

This is the **critical spinning speed** at which the bar passes from straight (zero deflection) to whirling motion when **one end** are prevented from bending and deflecting on long bearing while the **other end** on short bearing..

Case # 4: Rod supported on a long bearing at one end with other end free

Vanishing **deflection** and **tangent inclination** at fixed end and vanishing **bending moment** and **shear force** at the free end.

$$y(0) = 0 \qquad \text{vanishing deflection}$$
$$\left.\frac{dy}{dx}\right|_{x=0} = 0 \qquad \text{vanishing rotation} \tag{7-9.1a}$$

$$\left.\frac{d^2y}{dx^2}\right|_{x=l} = 0 \qquad \text{vanishing moment of bending}$$
$$\left.\frac{d^3y}{dx^3}\right|_{x=l} = 0 \qquad \text{vanishing shear force} \tag{7-9.1b}$$

Where, the shear force is given by equation (7-5.3) as

$$F = \frac{dM}{dx} = \frac{d^3y}{dx^3} \tag{7-9.1c}$$

Substituting by the boundary conditions in the deflection equation (7-5.5), we get

$$y(x) = A + F = 0$$
$$y'(x) = q[B+C] = 0$$
$$y''(x) = q^2\left[-A\cos(ql) - B\sin(ql) + C\sinh(ql) + F\cosh(ql)\right] = 0 \tag{7-9.2a}$$
$$y'''(x) = q^3\left[+A\sin(ql) - B\cos(ql) + C\cosh(ql) + F\sinh(ql)\right] = 0$$

Thus,

$$F = -A \qquad (7\text{-}9.2b)$$
$$B = -C \qquad (7\text{-}9.2c)$$

Like in the previous example, substituting by those in the remaining two equations, and dividing the two equations by members, we get

$$(-\cos(q\,l) - \cosh(q\,l))(\cos(q\,l) + \cosh(q\,l)) = (\sinh(q\,l) + \sin(q\,l))(\sin(q\,l) - \sinh(q\,l))$$

i.e., $\qquad \cos(q\,l)\cosh(q\,l) = -1 \qquad\qquad (7\text{-}9.3)$

The roots of this equation could be shown graphically to approximate

$$q\,l = 1.87, \frac{3\pi}{2}, \frac{5\pi}{2}, \dots \qquad (7\text{-}9.4)$$

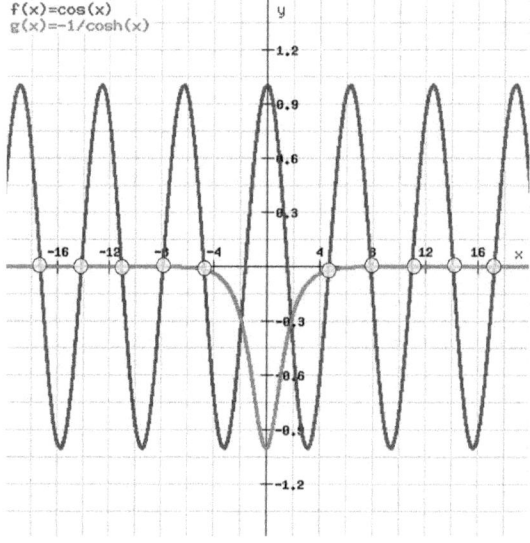

Figure 7-9. Roots of the transcendental equation (cos x cosh x = -1).

Substituting by the value of q_n from (7-9.4) in equation (7-5.4e), we get

$$\omega_1 = \frac{3.5}{l^2}\sqrt{\frac{gEJ}{W}}$$

$$\omega_{n+1} = \sqrt{\frac{gEJ}{W}}\left(\left(n+\frac{1}{2}\right)\frac{\pi}{l}\right)^2, \ (n = 1, 2, \ldots) \tag{7-9.5}$$

This is the **critical spinning speed** at which the bar passes from straight (zero deflection) to whirling motion when **one end** is prevented from bending and deflecting, while the other end is free.

7-2. Applications on P.D.E.

7-2.1. Wave equation of transverse vibration

The equation of oscillation in medium, without change in form of medium, is as follows

$$\overset{\cdots}{u} - k^2\frac{\partial^2 u}{\partial x^2} = 0 \tag{7-10.1a}$$

The dependent function u(x,t) represents displacement of the particles of the medium of elastic matter in case of **elastic vibration**, displacement of particles in case of **sound propagation**, displacement of surface particles in case of **surface waves**, and some unknown displacement in **electromagnetic waves**.

In case of elastic vibration, the speed of propagation of wave, k^2, is given by

$$k^2 = \frac{2\mu + \lambda}{\rho} \tag{7-10.2a}$$

The two constants μ and λ are elastic parameters of the medium known as **Lamé's coefficients**, defined by

$$\mu = \frac{E}{2(1+v)} \tag{7-10.2b}$$

$$\lambda = \frac{vE}{(1-2v)(1+v)} \tag{7-10.2c}$$

Where, E is the **Young's modulus**, v the **Poisson's ratio**, and ρ is the mass density of the medium

The **particular solution** of the wave equation (7-10.1a) takes the form;

283

$$u(x,t) = \sin(x \pm kt) \qquad (7\text{-}10.3)$$

The term $\pm kt$ determines the direction of motion of vibration from an arbitrary point x.

Example 101

Consider that case of simple one-dimensional case of transverse transmission of waves in a medium of **mass density** ρ exposed to external stress. Consider also that the wave is directed along the x direction of propagation.

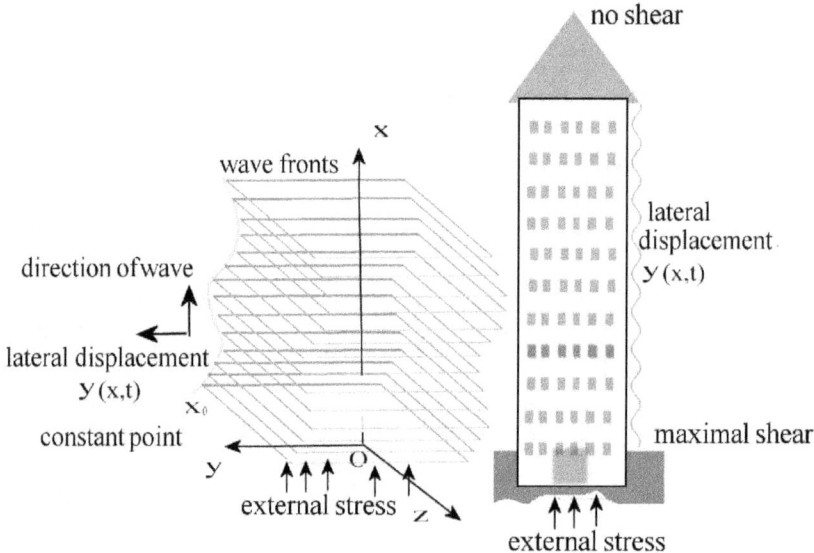

Figure 7-10. Transverse wave transmission in a vertical elastic body.

We need to investigate the shear waves comprising of **lateral displacements** y, perpendicular to direction x of propagation. Any such waves are determined by the shear elastic properties of the medium. **Hook's shear law** relates shear strain $\varepsilon(x,y,t)$, to shear stress, $\sigma(x,y,t)$, as follows

$$\sigma(x,y,t) = \mu\varepsilon(x,y,t) \qquad (7\text{-}11.1a)$$

And **Cauchy's relation** relates shear strain to displacement of deformation as follows

284

$$\varepsilon(x, y, t) = \frac{\partial y}{\partial x}$$

(7-11.1b)

The internal shear stress $\sigma(x,y,t)$ changes across infinitesimal distance dx by $\sigma + \frac{\partial \sigma}{\partial x} dx$.

The **equation of motion** of the infinitesimal element dx, cross section area ΔA, density ρ governed by **Newton's second** law, is

$$\left(\sigma + \frac{\partial \sigma}{\partial x} dx - \sigma\right)\Delta A = \rho \Delta A dx \frac{\partial^2 y}{\partial t^2}$$

(7-11.2a)

This could simplified to

$$\frac{\partial \sigma}{\partial x} = \rho \frac{\partial^2 y}{\partial t^2}$$

(7-11.2b)

This is **Navier's equation** in one dimension problem. On the LHS, we have the shear force of gradient of shear stress. On the RHS, we have the inertial acceleration induced by external stresses on the medium.

Navier's equation can also be written in terms of momentum and shear force as follows:

$$F = \frac{\partial \sigma}{\partial x} = \rho \frac{\partial^2 y}{\partial t^2} = \frac{\partial}{\partial t}\left(\rho \frac{\partial y}{\partial t}\right) = \frac{\partial M}{\partial t}$$

(7-11.2c)

Substituting by Hooke's law (7-11.1b) and Cauchy's relation (7-11.1a), we get

$$\frac{\partial}{\partial x}\left(\mu \frac{\partial y}{\partial x}\right) = \rho \frac{\partial^2 y}{\partial t^2}$$

(7-11.2c)

If we assume that, the shear modulus μ is independent of x, such as in homogeneous elastic medium, we get

$$\mu \frac{\partial^2 y}{\partial x^2} = \rho \frac{\partial^2 y}{\partial t^2}$$

(7-11.2d)

We have solved this equation by separation of variables in equation (6-6.7), as follows.

$$y(x, t) = X(x)T(t)$$
$$= \left(A\cos kat + B\sin kat\right)\left[C\cos kx + F\sin kx\right] \tag{7-11.3a}$$

By proper substitution in equation (6-6.7) with: $y = u$ and $a = \mu/\rho$, we get

$$y(x, t) = \left(A\cos\sqrt{\frac{\mu}{\rho}}kt + B\sin\sqrt{\frac{\mu}{\rho}}kt\right)\left[C\cos kx + F\sin kx\right] \tag{7-11.3b}$$

What remains is the determination of the four constants from the boundary conditions.

Boundary conditions

1. Vanishing lateral motion $y(0) = 0$ at $x = 0$ due to ground fixation

$$y(0, t) = \left(A\cos\sqrt{\frac{\mu}{\rho}}kt + B\sin\sqrt{\frac{\mu}{\rho}}kt\right)[C] = 0 \tag{7-11.4a}$$

Therefore, $C = 0$.

If we know the function $F(0,t)$ that describes the ground shaking we could substitute that either in the equation of shear stress at $x = 0$ or in the equation of displacement.

2. Vanishing shear stress $\dfrac{\partial y(x, t)}{\partial x}$, equations (7-11.1a and 1b), at top of building $x = h$

$$\frac{\partial y(x, t)}{\partial x} = \frac{\mu}{\rho}\left(A\cos\sqrt{\frac{\mu}{\rho}}kt + B\sin\sqrt{\frac{\mu}{\rho}}kt\right)\left[F\sin(kh)\right] = 0 \tag{7-11.4b}$$

Thus, as $\sin(kh) = 0$, the possible values of k are

$$k = \frac{n\pi}{h} \tag{7-11.4c}$$

Returning to the four unknown constants in equation (7-11.3a), we note that we have already determined two (C = 0 and $k = \dfrac{n\pi}{h}$) since F could now be included within A and B such that

$$y(x, t) = \sum_{n=1}^{n}\left(A_n\cos\sqrt{\frac{\mu}{\rho}}\frac{n\pi}{h}t + B_n\sin\sqrt{\frac{\mu}{\rho}}\frac{n\pi}{h}t\right)\sin\frac{n\pi}{h}x \tag{7-11.5a}$$

One of the remaining constants, namely A_n could be eliminated on the assumption that $y(x,0)$ vanishes at $t = 0$. Therefore, we have

$$y(x,t) = \sum_{n=1}^{n} B_n \sin \sqrt{\frac{\mu}{\rho} \frac{n\pi}{h}} t \sin \frac{n\pi}{h} x \qquad (7\text{-}11.5b)$$

We have excluded the case of $n - 0$ since such substitution will yield a single particular solution, which the reader could entertain finding and adding it to the general solution.

Kinetic energy of transverse vibration

For a building with cross section A, mass density per unit height ρ the kinetic energy is calculated by differentiating $y(x,t)$ with respect to time (speed of vibration wave), squaring the obtained speed, multiplying the squared speed with half the height density, and integrating over x from ground $x = 0$ to top, $x = h$, as following:

$$K.E. = \int \frac{1}{2} \rho A \left(\overset{"}{y}(x,t) \right)^2 dx$$

$$= \frac{1}{2} A\rho \left(\frac{\mu}{\rho} \frac{\pi^2}{h^2} \right) \int \left(\sum_{n=1}^{n} B_n n \cos \sqrt{\frac{\mu}{\rho} \frac{n\pi}{h}} t \sin \frac{n\pi}{h} x \right)^2 dx \qquad (7\text{-}11.6a)$$

We have prove before, in equation (6-7.10d), that

$$\int_{x=0}^{l} F_n \sin \frac{m\pi x}{l} \sin \frac{n\pi x}{l} dx = \frac{l}{2} F_n \qquad (7\text{-}11.6b)$$

Only if $m = n$, otherwise, the integral vanishes.

Therefore, equation (7-11.6a) become

$$K.E. = \frac{1}{2} A\mu \frac{\pi^2}{h^2} \left(\frac{h}{2} \right) \left(\sum_{n=1}^{n} B_n^2 n^2 \cos^2 \sqrt{\frac{\mu}{\rho} \frac{n\pi}{h}} t \right) \qquad (7\text{-}11.6c)$$

Or

$$K.E. = \frac{1}{4} A\mu \frac{\pi^2}{h} \left(\sum_{n=1}^{n} B_n^2 n^2 \cos^2 \sqrt{\frac{\mu}{\rho} \frac{n\pi}{h}} t \right) \qquad (7\text{-}11.6d)$$

Thus, at any instance t, **kinetic energy** of transverse vibration, of a building with height h, height-density ρ, shearing modulus μ, cross sectional area A, is the sum of infinite modes given by the above equation.

It should be emphasized that we have made great simplifications in the above example, which renders the results demonstrative rather than accurate. Among our crude assumption is the entire cross section of the building is subjected to uniform shearing stress, the height-density constant, the building has no internal forces such as its own weight, that transverse vibrations were pure, isometric and occurring in homogeneous medium.

7-2.2. General wave equation of vibration

A. Transverse vibration

(i) Equations of displacement:

In the previous example we discussed the simplest problem of one-dimensional transverse or lateral vibration where the direction of wave propagation was x, and the deformation of the medium was in the y-direction.

Let us consider the general notation of **three-dimensional vibration** as follows.

The medium exposed to external stress is deformed in the x-, y-, and z-directions by the displacements $u(x,y,z,t)$, $v(x,y,z,t)$, and $w(x,y,z,t)$,. This, transverse vibrations when the wave travels on the x-axis and displacements occurs only on the z, axis, are described by the general displacement equations:

$$u = 0$$
$$v = 0$$
$$w = w(x,t) \qquad\qquad (7\text{-}12.1)$$

(ii) Body forces: assumed absent

$$X_i = Y_i = Z_i = 0 \qquad\qquad (7\text{-}12.2)$$

(iii) Principal strain from Cauchy's equations

Since the displacement w occurs in the transverse yz-plane, then all points will move equally causing no strain in the three directions. Therefore,

$$\varepsilon = \varepsilon_{xx} + \varepsilon_{yy} + \varepsilon_{zz} = \frac{\partial u}{\partial x} + \frac{\partial v}{\partial y} + \frac{\partial w}{\partial z} = 0 \qquad\qquad (7\text{-}12.3)$$

This equation is known in the **theory of elasticity** as the equation of the principal dilatational strain. It implies that the principal axis of a medium stressed within its elastic capacity is defined by vanishing net strain during vibration alone.

The three derivatives of the principal strain ε also vanish due to absent stresses on the principal axis. Thus,

288

$$\frac{\partial \varepsilon}{\partial x} = \frac{\partial^2 u}{\partial x^2} = 0 \qquad \frac{\partial \varepsilon}{\partial y} = \frac{\partial^2 u}{\partial x \partial y} = 0 \qquad \frac{\partial \varepsilon}{\partial z} = \frac{\partial^2 u}{\partial x \partial z} = 0 \qquad (7\text{-}12.4)$$

The Laplacians of equations (7-12.1) are obtained from (7-12.4) and the acceleration terms along the y- and z-axes vanish, from (7-12.1). Thus

$$\nabla^2 u = 0 \qquad \nabla^2 v = 0 \qquad \nabla^2 w = \frac{\partial^2 w}{\partial x^2} \qquad \frac{\partial^2 v}{\partial t^2} = \frac{\partial^2 u}{\partial t^2} = 0 \qquad (7\text{-}12.5)$$

Note that w(x,t) can only change along the x-direction.

(iv) Wave equation of transverse vibration

$$\mu \frac{\partial^2 w}{\partial x^2} = \rho \frac{\partial^2 w}{\partial t^2} \qquad (7\text{-}12.6)$$

The **shearing modulus** μ, often denoted by G, is assumed constant for homogeneous isotropic medium. The same is assumed for the medium mass density ρ.

B. General wave equation

Vibrational motions in elastic medium treated above are special case of **the Lamé's equations**, that involve the spatial derivatives of displacements and dilatational strain weighed with the Lamé's coefficients, internal body forces, and external inertial forces as follows

$$\mu \nabla^2 \begin{bmatrix} u \\ v \\ w \end{bmatrix} + (\mu + \lambda) \begin{bmatrix} \partial \varepsilon / \partial x \\ \partial \varepsilon / \partial y \\ \partial \varepsilon / \partial z \end{bmatrix} + \rho \begin{bmatrix} X_i \\ Y_i \\ Z_i \end{bmatrix} = \rho \frac{\partial^2}{\partial t^2} \begin{bmatrix} u \\ v \\ w \end{bmatrix} \qquad (7\text{-}13.1)$$

This is a typical three wave-equations matrix in three-dimensional homogenous medium. The internal forces X_i, Y_i, and Z_i are assumed per unit mass and the two **Lamé's coefficients** μ and λ arising from the application of **Hooke's volumetric law** of stress-strain in homogenous isotropic elastic medium.

(i) Dilatational wave equation

In order to express equation (7-13.1) in terms of **dilatational strain**, we will differentiate each of the three equations with respect to each respective coordinate and then add the three equations to produce a single partial differential equation as follows

$$\mu\nabla^2\begin{bmatrix}\partial u/\partial x\\\partial v/\partial y\\\partial w/\partial z\end{bmatrix}+(\mu+\lambda)\begin{bmatrix}\partial^2\varepsilon/\partial x^2\\\partial^2\varepsilon/\partial y^2\\\partial\varepsilon^2/\partial z^2\end{bmatrix}=\rho\frac{\partial^2}{\partial t^2}\begin{bmatrix}\partial u/\partial x\\\partial v/\partial y\\\partial w/\partial z\end{bmatrix}\qquad(7\text{-}13.2)$$

Adding the three equations

$$\mu\nabla^2\left[\frac{\partial u}{\partial x}+\frac{\partial v}{\partial y}+\frac{\partial w}{\partial z}\right]+(\mu+\lambda)\nabla^2\varepsilon=\rho\frac{\partial^2}{\partial t^2}\left[\frac{\partial u}{\partial x}+\frac{\partial v}{\partial y}+\frac{\partial w}{\partial z}\right]\qquad(7\text{-}13.3)$$

Where the **divergence of displacements** (the bracketed sums of derivatives) comprises the **dilatational strain**, thus simplifying the wave equation to the following

$$(2\mu+\lambda)\nabla^2\varepsilon=\rho\frac{\partial^2\varepsilon}{\partial t^2}\qquad(7\text{-}13.4)$$

Where the **principal strain** ε is defined in equation (7-12.3).

The velocity of the displacement wave is defined (see also equation 7-10.2a)

$$c_L=\pm\sqrt{\frac{2\mu+\lambda}{\rho}}\qquad(7\text{-}13.5)$$

Equation (7-13.4) can be written in more representative form in terms of a **displacement vector** **u**(u,v,w) as follows

$$\mu\nabla^2[\nabla.\mathbf{u}]+(\mu+\lambda)\nabla^2\varepsilon=\rho\frac{\partial^2}{\partial t^2}[\nabla.\mathbf{u}]\qquad(7\text{-}13.6)$$

In case that the reader is unfamiliar with the theory of elasticity, in the latter, strain is defined by **Cauchy's equations** as the gradient of directional displacements (u, v, w in the directions of x, y, z) along those direction. In simple words, strain is displacement per unit distance.

(ii) Equivoluminal S-waves

If the dilatational strain ε vanishes, then equation (7-13.6) becomes

$$\mu\nabla^2[\nabla.\mathbf{u}]\varepsilon=\rho\frac{\partial^2}{\partial t^2}[\nabla.\mathbf{u}]\qquad(7\text{-}13.7)$$

Which represents vibrations in displacements with net vanishing dilatation. Thus, the speed of propagation is mainly of distortion or rotation or shear strains (S-waves) given by

$$c_S = \pm\sqrt{\frac{\mu}{\rho}} \qquad (7\text{-}13.8)$$

Recall that equation (7-13.7) is one among three wave equation, the remaining two are for v and w.

(iii) Distortional or solenoidal wave equation

In order to derive the **rotational or solenoidal wave equation**, we differentiate the second equation with respect to z, the third with respect to y, we get

$$\mu\nabla^2\begin{bmatrix}\partial v/\partial z \\ \partial w/\partial y\end{bmatrix} + (\mu+\lambda)\begin{bmatrix}\partial^2\varepsilon/\partial y\partial z \\ \partial^2\varepsilon/\partial y\partial z\end{bmatrix} = \rho\frac{\partial^2}{\partial t^2}\begin{bmatrix}\partial v/\partial z \\ \partial w/\partial y\end{bmatrix} \qquad (7\text{-}13.9)$$

Subtracting the two equations by member, we get

$$\mu\nabla^2\left(\frac{\partial w}{\partial y} - \frac{\partial v}{\partial z}\right) + (\mu+\lambda)\left[\frac{\partial^2\varepsilon}{\partial y\partial z} - \frac{\partial^2\varepsilon}{\partial y\partial z}\right] = \rho\frac{\partial^2}{\partial t^2}\left(\frac{\partial w}{\partial y} - \frac{\partial v}{\partial z}\right) \qquad (7\text{-}13.10)$$

Replacing the bracketed differences of derivatives by **rotational strains**, we get

$$\mu\nabla^2\omega_x = \rho\frac{\partial^2\omega_x}{\partial t^2} \qquad (7\text{-}13.11)$$

Rotational strains are defined in the theory of elasticity as the components **curl** of the **displacement vector**. Thus, rotational strain along the x-axis is represented by the difference bf the gradients of the two perpendicular displacement projections (v and w) and is denotes by ω_x.

Similarly, we can obtain the remaining two equations for y- and z- dependent rotational strains

$$\mu\nabla^2\omega_x = \rho\frac{\partial^2\omega_x}{\partial t^2}$$
$$\mu\nabla^2\omega_y = \rho\frac{\partial^2\omega_y}{\partial t^2} \qquad (7\text{-}13.12)$$
$$\mu\nabla^2\omega_z = \rho\frac{\partial^2\omega_z}{\partial t^2}$$

The velocity of the **rotational waves** is defined as follows

$$c_S = \pm\sqrt{\frac{\mu}{\rho}} \qquad (7\text{-}13.13)$$

Equations (7-13.12) is also put into more representative form in terms of a **solenoid vector ω** $(ω_x, ω_y, ω_z)$

$$\mu\nabla^2(\nabla\mathbf{x}\mathbf{u}) = \rho\frac{\partial^2}{\partial t^2}(\nabla\mathbf{x}\mathbf{u})$$ (7-13.14)

Where

$$\boldsymbol{\omega}(\omega_x, \omega_y, \omega_z) = \nabla\mathbf{x}\mathbf{u}$$ (7-13.15)

Equation (7-13.14) is called the wave equation for:

1. Rotational waves
2. Equivoluminal waves
3. Distortional waves
4. Shear waves
5. Transverse waves
6. S-waves

(iv) Irrotational or P-wave equation

The conditions of non-rotating elastic vibration waves comprise of vanishing curling components of stresses as follows;

$$\frac{\partial w}{\partial y} - \frac{\partial v}{\partial z} = 0$$
$$\frac{\partial w}{\partial x} - \frac{\partial u}{\partial z} = 0$$ (7-14.1)
$$\frac{\partial u}{\partial y} - \frac{\partial v}{\partial x} = 0$$

The reader might question the causation of vanishing curling stresses in an elastic medium. Such conditions are caused by the distribution of **external stresses** and **surface conditions** of the medium. Those two factors entail all parameters pertaining to the direction, magnitude, and spatial and temporal profiles of the external stresses and the surface geometry of the medium that allows internal stresses to be coupled to external stresses in many ways, such as creating irrotational waves, Figure 7-11.

292

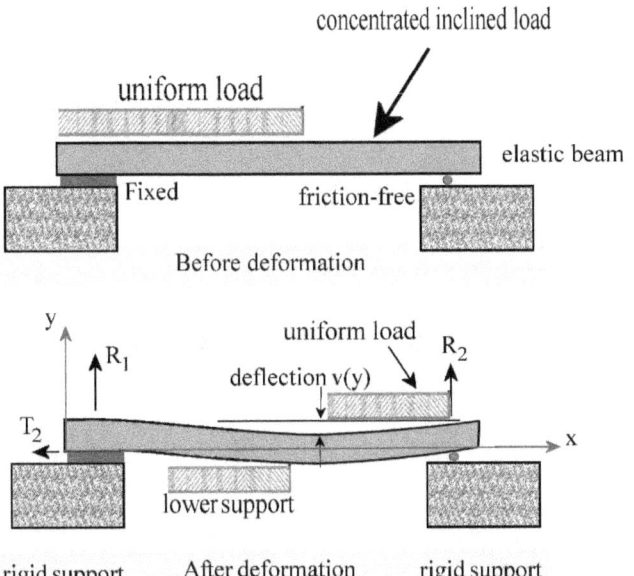

Figure 7-11. The role of surface conditions (loading conditions) and external load profiles on generating internal elastic stresses.

Therefore, we could assume that the three displacements u, v, and w are derivatives of the same arbitrary potential Φ, as follows

$$u = \frac{\partial \Phi}{\partial x}, \qquad v = \frac{\partial \Phi}{\partial y}, \qquad w = \frac{\partial \Phi}{\partial z} \qquad (7\text{-}14.2)$$

Thus, the potential dilatational strain becomes

$$\begin{aligned}
\varepsilon &= \frac{\partial u}{\partial x} + \frac{\partial v}{\partial y} + \frac{\partial w}{\partial z} \\
&= \frac{\partial^2 \Phi}{\partial x^2} + \frac{\partial^2 \Phi}{\partial y^2} + \frac{\partial^2 \Phi}{\partial z^2} \qquad (7\text{-}14.3) \\
&= \nabla^2 \Phi
\end{aligned}$$

From, equations (7-14.2) and (7-14.3), the derivatives of the dilatational strain can be represented as

293

$$\frac{\partial \varepsilon}{\partial x} = \frac{\partial}{\partial x}\left(\nabla^2 \Phi\right)$$

$$= \nabla^2 \left(\frac{\partial \Phi}{\partial x}\right) \tag{7-14.4}$$

$$= \nabla^2 u$$

Or,

$$\frac{\partial \varepsilon}{\partial x} = \nabla^2 u, \qquad \frac{\partial \varepsilon}{\partial y} = \nabla^2 v, \qquad \frac{\partial \varepsilon}{\partial z} = \nabla^2 w \tag{7-14.5}$$

Substituting from equations (7-14.5) into (7-13.1), the **potential wave equation** becomes

$$\mu \nabla^2 \begin{bmatrix} u \\ v \\ w \end{bmatrix} + (\mu + \lambda)\nabla^2 \begin{bmatrix} u \\ v \\ w \end{bmatrix} = \rho \frac{\partial^2}{\partial t^2} \begin{bmatrix} u \\ v \\ w \end{bmatrix} \tag{7-14.6}$$

Or

$$(2\mu + \lambda)\nabla^2 \begin{bmatrix} u \\ v \\ w \end{bmatrix} = \rho \frac{\partial^2}{\partial t^2} \begin{bmatrix} u \\ v \\ w \end{bmatrix} \tag{7-14.7}$$

Which is a wave equation for displacements, given any of the following names:

1. Longitudinal waves
2. P-waves
3. Bulk waves
4. Dilatational waves
5. Irrotational waves

(v) Spherical wave equation

Consider the general form of wave equation

$$\frac{\partial^2 f(x, y, z, t)}{\partial t^2} = c^2 \nabla^2 f(x, y, z, t) \tag{7-15.1}$$

Then, consider the general solution for equation (7-15.1) in the form

$$f = f_o(x, y, z, t, \alpha, \beta, \xi, \ldots) \tag{7-15.2}$$

Differentiate equation (7-15.1) with respect to any of the arguments of f_0, as follows

$$\frac{\partial}{\partial \xi}\left(\frac{\partial^2 f_0}{\partial t^2}\right) = c^2 \frac{\partial}{\partial \xi}\left(\nabla^2 f_0\right)$$

$$\frac{\partial^2}{\partial t^2}\left(\frac{\partial f_0}{\partial \xi}\right) = c^2 \nabla^2 \left(\frac{\partial f_0}{\partial \xi}\right)$$

(7-15.3)

Therefore, we conclude that the **derivatives of the wave function** with respect to its derivatives are also solutions of the wave equation.

The simple case of **spherical wave** can be represented by the **radial** and **temporal** coordinates of the form

$$f = f_0(r,t)$$

where,

$$r = \sqrt{x^2 + y^2 + z^2}$$

(7-15.4)

The derivatives of f_0 are obtained as follows

$$\frac{\partial r}{\partial x} = \frac{x}{r}$$

$$\frac{\partial f_0}{\partial x} = \frac{\partial f_0}{\partial r}\frac{\partial r}{\partial x} = \frac{x}{r}\frac{\partial f_0}{\partial r}$$

(7-15.5)

Thus, the second derivatives with respect to x, and similarly, for y- and z- derivatives, are

$$\frac{\partial^2 f_0}{\partial x^2} = \frac{x^2}{r^2}\frac{\partial^2 f_0}{\partial r^2} + \frac{r^2 - x^2}{r^3}\frac{\partial f_0}{\partial r}$$

$$\frac{\partial^2 f_0}{\partial y^2} = \frac{y^2}{r^2}\frac{\partial^2 f_0}{\partial r^2} + \frac{r^2 - y^2}{r^3}\frac{\partial f_0}{\partial r}$$

$$\frac{\partial^2 f_0}{\partial z^2} = \frac{z^2}{r^2}\frac{\partial^2 f_0}{\partial r^2} + \frac{r^2 - z^2}{r^3}\frac{\partial f_0}{\partial r}$$

(7-15.6)

Adding equations (7-15.6) by members we get

$$\nabla^2 f_0 = \frac{\partial^2 f_0}{\partial r^2} + \frac{2}{r}\frac{\partial f_0}{\partial r}$$

$$= \frac{1}{r}\frac{\partial^2}{\partial r^2}(rf_0)$$

(7-15.7)

Thus, from equations (7-15.1) and (7-15.7), the spherical wave equation becomes

$$\frac{\partial^2 f_0}{\partial t^2} = \frac{c^2}{r} \frac{\partial^2}{\partial r^2}(rf_0) \qquad (7\text{-}15.8)$$

Since r is an independent variable, equation (7-15.8) can be written as

$$\frac{\partial^2}{\partial t^2}(rf_0) = c^2 \frac{\partial^2}{\partial r^2}(rf_0) \qquad (7\text{-}15.9)$$

Equation (7-15.9) can be solved by separation of variables as follows

$$f_0 = R(r)T(t) \qquad (7\text{-}15.10)$$

Such that

$$RT'' = c^2 TR''$$

$$(7\text{-}15.11)$$

$$\text{Or} \qquad \frac{T''}{T} = c^2 \frac{R''}{R}$$

The two separate equation in R and T, are written as usual in the form

$$T' + \lambda^2 T = 0$$

$$R'' + \frac{\lambda^2}{c^2} R = 0 \qquad (7\text{-}15.12)$$

Those are second-order o.d.e. which we have solved extensively through the previous chapters of this book.

7-2.3. Plane sound waves

(a) Deformations of volume, density, and pressure

Example 102

Let us consider sound wave propagation in gas medium.

Initially, $t = 0$, an infinitesimal volume of cross sectional are ΔA perpendicular to the direction of wave propagation x, and with longitudinal dimension δx, Figure 7-12, is given by

$$V_o = (\Delta A)(\delta x) \tag{7-16.1}$$

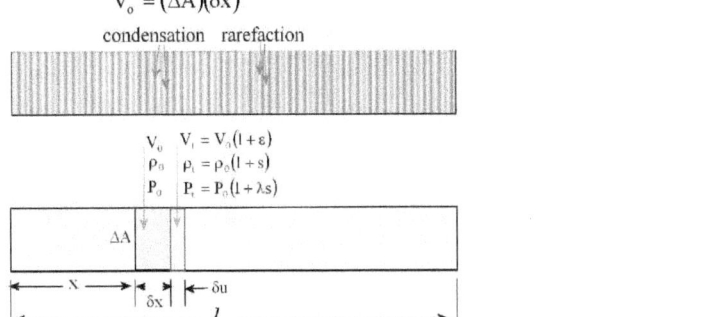

Figure 7-12. Propagation of sound waves in gas medium.

As the sound propagates in the medium, $t = \tau$, the infinitesimal volume changes to

$$V_t = (\Delta A)(\delta x + \delta u) \tag{7-16.2}$$

Where, u is the displacement of gas particles due to sound waves' **compression/rarefaction** stresses.

Equation (7-16.2) can be written as

$$V_t = (\Delta A)\delta x \left(1 + \frac{\delta u}{\delta x}\right) \tag{7-16.3}$$

The differential fraction of displacement over distance is called **dilatational strain** denoted by ε. Hence,

$$V_t = V_0(1 + \varepsilon) \tag{7-16.4}$$

We have thus reduced the problem of sound wave propagation into a problem of elastic deformation of the medium where particles are strained infinitesimally. The second feature of all elastic problems (**strain** being the first feature) is to find the **stress** that induced such strain.

Similarly, the density ρ of gas particles changes as sound pressure waves propagates, as follows

$$\rho_t = \rho_0(1 + s) \tag{7-16.5}$$

Where s is the **condensation strain**.

On the assumption that sound waves do not create mass defect, mass is conserved before and after the passage of sound waves. The conservation of mass can be written as follows

$$\rho_t V_t = \rho_0 V_0 \tag{7-16.6}$$

From equation (7-16.4) and (7-16.5), we got

$$\rho_0 V_0 (1+s)(1+\varepsilon) = \rho_0 V_0$$
$$(1+s)(1+\varepsilon) = 1 \tag{7-16.7}$$

Since dilatational strain ε amounts to rarefaction or **diminution in mass density**, we could relate s to ε as follows

$$\varepsilon = -s = \frac{\partial u}{\partial x} \tag{7-16.8}$$

We now need to find the force behind causing the change of momentum of gas particle. For that, we use the law of perfect gases stated as follows

$$P_t \rho_t{}^\lambda = P_0 \rho_0{}^\lambda = nR \tag{7-16.9}$$

The constant λ characterizes each different gas, n is the particle density, and R is the universal gas constant.

Equation (7-16.9) holds true when there is mutual transfer of heat between adjacent gas particles. In the theory of elasticity, this approximation is called **plane-strain** or **plane-stress** approximation.

Thus, a pressure wave similar to volume wave (7-16.4) and density wave (7-16.6) is as follows

$$\rho_t{}^\lambda = \rho_0{}^\lambda (1+s)^\lambda$$
$$P_t = P_0 (1+s)^\lambda \tag{7-16.10}$$

Since, s is small, we could expand the power term in binomial series and ignore high powers of s to get

$$P_t = P_0 (1+\lambda s) \tag{7-16.11}$$

(b) Equation of motion

The **force** imposed on the medium is the product of pressure and cross-sectional area as follows

298

$$F = -P_t(\Delta A) \qquad (7\text{-}16.12a)$$

Here, the negative sign signifies rarefaction, positive signifies condensation.

From (7-16.11), the force is

$$F = -P_0(1 + \lambda s)(\Delta A) \qquad (7\text{-}16.12b)$$

Substituting from (7-16.8), we get

$$F = -P_0\left(1 - \lambda \frac{\partial u}{\partial x}\right)(\Delta A) \qquad (7\text{-}16.12c)$$

The **momentum** gained by the particles of the medium is product of mass (density times the cross section area times the length) and velocity. This is also defined by our already described equations, (7-16.3), as

$$M_t = \rho_0(\Delta A)\frac{\partial}{\partial t}\left(\delta x\left(1 + \frac{\partial u}{\partial x}\right)\right) \qquad (7\text{-}16.13a)$$

Since δx is assumed the constant length of the infinitesimal volume, we will use the differential property

$$\delta x \frac{\partial u}{\partial x} = \delta u$$
$$\frac{\partial}{\partial t}\delta u = \frac{\partial u}{\partial t} \qquad (7\text{-}16.13b)$$

Thus, the momentum of the infinitesimal element in (7-16.13a) becomes

$$M_t = \rho_0(\Delta A)\frac{\partial u}{\partial t} \qquad (7\text{-}16.13c)$$

Newton's second law of motion requires that the **loss** of force upon propagation in the medium is transferred to **gain** in momentum of the medium particles.

This is formulated mathematically as follows.

Change in force, from equation (7-16.12c), is

$$\frac{\partial F}{\partial x} = P_0 \lambda \frac{\partial^2 u}{\partial x^2} (\Delta A)$$

(7-16.14a)

Change in momentum, from equation (7-16.13c), is

$$\frac{\partial M_t}{\partial t} = \rho_0 (\Delta A) \frac{\partial^2 u}{\partial t^2}$$

(7-16.14b)

The **equation of motion** comprises the balance between gradient of force over space (distance) and rate of change of momentum over time. Thus, equations (7-16.14a and 14b) give

$$\frac{\partial^2 u}{\partial t^2} = \frac{P_0 \lambda}{\rho_0} \frac{\partial^2 u}{\partial x^2}$$

(7-16.15a)

This is the wave equation of sound propagation in gas with speed given by

$$c = \sqrt{\frac{P_0 \lambda}{\rho_0}}$$

(7-16.15b)

The solution of the wave equation is subjected to the boundary conditions of any given problem.

For example, a **bar of length** l, should have the following boundary conditions:

$$\varepsilon = -s = \frac{\partial u}{\partial x} = 0 \text{ at } x = 0 \text{ and } x = l.$$

(7-16.16a)

The separate variable equations of the sound wave, (7-16.15a), are

$$\frac{1}{c^2 T} \frac{d^2 T}{dt^2} = \frac{1}{X} \frac{d^2 X}{dx^2} = k^2$$

$$\frac{d^2 T}{dt^2} - Tc^2 k^2 = 0$$

$$\frac{d^2 X}{dx^2} = Xk^2 = 0$$

(7-16.16b)

Yielding the solution for longitudinal vibration as

$$u(x, t) = \sum_{n=0}^{\infty} C_n \cos \frac{n\pi x}{l} \cos \frac{n\pi c t}{l}$$

(7-16.16c)

Using the trigonometric property of cosines:

$$\cos(a+b) = \cos a \cos b - \sin a \sin b$$
$$\cos(a-b) = \cos a \cos b + \sin a \sin b \qquad (7\text{-}16.16d)$$
$$\cos a \cos b = \cos(a+b) + \cos(a-b)$$

Equation (7-16.16c) can be written in a more meaningful form as follows

$$u(x,t) = \sum_{n=0}^{\infty} C_n \left[\cos\left(\frac{n\pi}{l}(x+ct)\right) + \cos\left(\frac{n\pi}{l}(x-ct)\right) \right] \qquad (7\text{-}16.16e)$$

Symbolically, we refer to the above wave equation by the following notation

$$u(x,t) = \varphi(x+ct) + \varphi(x-ct) \qquad (7\text{-}16.17)$$

This signifies the propagation of two waves, in opposite directions, one of condensation, the other of rarefaction, such that the medium incurs no permanent deformation, while transmitting sound waves.

7-2.4. Electric waves in a conductor

Example 103

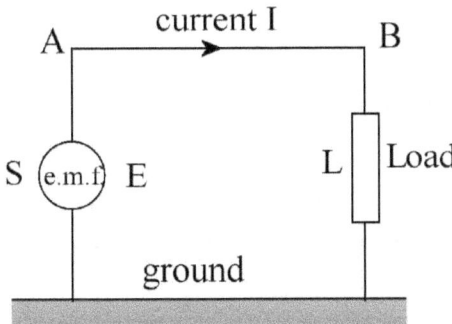

Figure 7-13. Flow of electricity in wire.

Consider insulated conductor AB, energized between A and the ground with the electric source S (volts) and loaded between B and the ground with load L.

At any point in time t and place x, the electric current, i (amperes) and potential e (volt) will vary according to the nature of load and intermediate leakage sources in the line.

301

We could now restate equation (7-1.5) such that the voltage drops across the various electric elements depend on their lengths. We first need to separate the capacitive elements, which store charges from the resistive and inductive elements as follows.

$$\delta e_r = -\delta V_L - \delta V_R$$

$$= L\delta x \frac{\partial i}{\partial t} + (Ri)\delta x \qquad (7\text{-}17.1a)$$

$$\delta e_c = -\frac{1}{C}\int_0^t idt \qquad (7\text{-}17.1b)$$

We shall return later to show that the capacitive term in (7-17.1b) in affects the current in (7-17.1a).

These can be written as

$$-\frac{\partial e}{\partial x} = L\frac{\partial i}{\partial t} + Ri \qquad (7\text{-}17.2a)$$

$$-\frac{\partial e}{\partial t} = \frac{i}{C} \qquad (7\text{-}17.2b)$$

The capacitive equation is written as

$$i = -C\frac{\partial e}{\partial t} \qquad (7\text{-}17.2c)$$

This allows us to introduce the infinitesimal changes as follows

$$\delta i = -C\delta x\frac{\partial e}{\partial t} \qquad (7\text{-}17.2d)$$

Which leads us to the differential equation for current across the capacitive elements of the circuit as follows

$$\frac{\partial i}{\partial x} = -C\frac{\partial e}{\partial t} \qquad (7\text{-}17.2e)$$

We could also expand the last equation to include voltage leaks such as Ge, where is a leakage constant of the line. Thus

$$\frac{\partial i}{\partial x} = -C\frac{\partial e}{\partial t} - Ge \qquad (7\text{-}17.2f)$$

Since we have treated the capacitive term in (7-17.2b) separate from the remaining voltage drops, we return to equation (7-17.2a), differentiate it with respect to x, in order to account for the hidden effect of the capacitive term, which we have redundantly assumed to affect the current.

Thus, equation (7-17.2a) upon differentiation gives

$$-\frac{\partial^2 e}{\partial x^2} = L\frac{\partial^2 i}{\partial t \partial x} + R\frac{\partial i}{\partial x} \tag{7-17.3a}$$

Substituting from equation (7-12.2f), we get

$$-\frac{\partial^2 e}{\partial x^2} = L\frac{\partial}{\partial t}\left(-C\frac{\partial e}{\partial t} - Ge\right) + R\left(-C\frac{\partial e}{\partial t} - Ge\right) \tag{7-17.3b}$$

Executing the differentiation, we get

$$\frac{\partial^2 e}{\partial x^2} = CL\frac{\partial^2 e}{\partial t^2} + (GL + CR)\frac{\partial e}{\partial t} + RGe \tag{7-17.3c}$$

This is the wave equation for the **propagation of energy or voltage** in a conductor.

We could get a similar equation for the current by differentiating (7-17.2a) with respect to t and get

$$-\frac{\partial^2 e}{\partial x \partial t} = L\frac{\partial^2 i}{\partial t^2} + R\frac{\partial i}{\partial t} \tag{7-17.4a}$$

Then, equation (7-17.2f) is used to eliminate the voltage term in the above equation as follows:

$$\frac{1}{C}\frac{\partial}{\partial x}\left(\frac{\partial i}{\partial x} + Ge\right) = L\frac{\partial^2 i}{\partial t^2} + R\frac{\partial i}{\partial t} \tag{7-17.4b}$$

Performing the differentiation, we get

$$\left(\frac{\partial^2 i}{\partial x^2} + G\frac{\partial e}{\partial x}\right) = CL\frac{\partial^2 i}{\partial t^2} + CR\frac{\partial i}{\partial t} \tag{7-17.4c}$$

The last substitution from (7-17.2a) eliminates the gradient of e, as follows

$$\left(\frac{\partial^2 i}{\partial x^2} - GL\frac{\partial i}{\partial t} - GRi\right) = CL\frac{\partial^2 i}{\partial t^2} + CR\frac{\partial i}{\partial t} \tag{7-17.4d}$$

303

Arranging, we get

$$\frac{\partial^2 i}{\partial x^2} = CL\frac{\partial^2 i}{\partial t^2} + (CR + GL)\frac{\partial i}{\partial t} + GRi \qquad\qquad (7\text{-}17.4e)$$

This is the wave equation for the **propagation of current** in a conductor.

Summary of wave propagation in electric conductors

Telephone Equations

$$-\frac{\partial e}{\partial x} = L\frac{\partial i}{\partial t} + Ri \qquad\qquad (7\text{-}17.2a)$$

$$\frac{\partial i}{\partial x} = -C\frac{\partial e}{\partial t} - Ge \qquad\qquad (7\text{-}17.2f)$$

$$\frac{\partial^2 e}{\partial x^2} = CL\frac{\partial^2 e}{\partial t^2} + (GL + CR)\frac{\partial e}{\partial t} + RGe \qquad\qquad (7\text{-}17.3c)$$

$$\frac{\partial^2 i}{\partial x^2} = CL\frac{\partial^2 i}{\partial t^2} + (CR + GL)\frac{\partial i}{\partial t} + GRi \qquad\qquad (7\text{-}17.4e)$$

The above four equations are known as Telephone Equations.

The **speed of the propagation of electrons** in conductor is:

$$c = \frac{1}{\sqrt{CL}} \qquad\qquad (7\text{-}17.5)$$

Example 104

Given the case of **Telegraph Equations** defined by negligible L (inductance) and G (leakage), in the **Telephone equations** such that

Telegraph Equations:

$$-\frac{\partial e}{\partial x} = Ri \qquad\qquad (7\text{-}18.1)$$

$$\frac{\partial i}{\partial x} = -C\frac{\partial e}{\partial t} \qquad\qquad (7\text{-}18.2)$$

$$\frac{\partial^2 e}{\partial x^2} = (CR)\frac{\partial e}{\partial t} \qquad\qquad (7\text{-}18.3)$$

$$\frac{\partial^2 i}{\partial x^2} = (CR)\frac{\partial i}{\partial t} \qquad\qquad (7\text{-}18.4)$$

Those are to be solved for a line of 100 km in length.
One end of the line is kept at 2 volts.
Other end kept at 6 volts until a steady state is reached.
The first end of 2 volt is suddenly grounded, while the other end is maintained at 6 volts.

$R = 0.12$ ohm / km
$C = 2$ microfarads /km

Find the steady-state solution.

Solution

$$RC = 0.12 \times 2 \times 10^{-6} \qquad\qquad (7\text{-}18.5)$$

The **steady state** solution for equation (7-18.3c) gives first-degree expression in x, as follows:

$$\frac{\partial^2 e}{\partial x^2} = 0$$

$$\frac{\partial e}{\partial x} = C \qquad\qquad (7\text{-}18.6)$$

$$e = Cx + B$$

At $x = 0$, $e = 2$ volts, thus $B = 2$.
At $x = 100$, $e = 6$, thus $6 = 100C+2$, or $C=4/100$

Thus, equation (7-18.6) gives the steady-state condition **prior to grounding** as

$$e(x,0) = \frac{x}{25} + 2 \qquad\qquad (7\text{-}18.7)$$

After grounding the first end, the boundary conditions become.

$$e(0,t) = 0$$
$$e(100, t) = 6 \qquad\qquad (7\text{-}18.8)$$

This correspond to the permanent steady-state condition, where at $x = 0$, $e(0,t)=0$ and $B = 0$. And
At $x = 100$, $e = 6$, thus $C = 6/100$.

Hence, the permanent steady-state after grounding is

$$e_s(x,0) = \frac{3x}{50} \qquad\qquad (7\text{-}18.9)$$

Equation (7-18.3c) must satisfy those both steady-states (7-18.7) and (7-18.9).

305

Thus, the **transient voltage** represents the difference between the two steady states as follows

$$e_T(x,0) = \left(\frac{x}{25} + 2\right) - \left(\frac{3x}{50}\right)$$

Or

$$e_T(x,0) = 2 - \frac{x}{50} \qquad (7\text{-}18.10)$$

We need to determine e_{Ts}, which comprises the transient voltage between the two steady states.

Solving equation (7=18.3c) by separation of variables, we get

$$\frac{1}{X}\frac{d^2X}{dx^2} = (CR)\frac{1}{T}\frac{dT}{dt} = -k^2$$

$$\frac{d^2X}{dx^2} + Xk^2 = 0 \qquad (7\text{-}18.11)$$

$$\frac{dT}{dt} + T\frac{k^2}{(CR)} = 0$$

$$e = \left(A\sin kx + B\cos kx\right)e^{-\frac{k^2}{(CR)}t} \qquad (7\text{-}18.12)$$

Substituting by the boundary conditions after grounding the first end (7-18.8), we get:

At $x = 0$:

$$\left(B\right)e^{-\frac{k^2}{(CR)}t} = 0 \quad \text{Or } B = 0 \qquad (7\text{-}18.13a)$$

At $x=100$, the transient voltage vanishes by equation (7-18.10), such that

$$0 = \left(A\sin k100\right)e^{-\frac{k^2}{(CR)}t}$$

Or

$$k100 = n\pi \qquad (7\text{-}18.13b)$$

Thus, the transient voltage, (7-18.12), is

$$e_T = \sum_{n=1}^{\infty}\left(A_n\sin\frac{n\pi x}{100}\right)e^{-\frac{1}{(CR)}\left(\frac{n\pi}{100}\right)^2 t} \qquad (7\text{-}18.14)$$

Equation (7-18.14) must also satisfy the initial condition (7-18.10) at $t = 0$ such that

306

$$\sum_{n=1}^{\infty} \left(A_n \sin \frac{n\pi x}{100} \right) = 2 - \frac{x}{50} \tag{7-18.15}$$

In order to determine A_n, we need to expand the RHS in Fourier's series that enable us to use Euler's formula to determine A_n.

$$\sum_{n=1}^{\infty} \left(A_n \int \sin \frac{m\pi x}{100} \sin \frac{n\pi x}{100} \right) dx = \int \left(2 - \frac{x}{50} \right) \sin \frac{m\pi x}{100} dx$$

$$= \frac{100}{m\pi} \int \left(2 - \frac{x}{50} \right) \sin \beta d\beta$$

$$= \frac{100}{m\pi} \left(4 - \frac{1}{50} \frac{100}{m\pi} \int \beta \sin \beta d\beta \right) \tag{7-18.16a}$$

$$= \frac{200}{m\pi}$$

And by virtue of equation (6-7.10d), we have

$$\int_{x=0}^{l} F_n \sin \frac{m\pi x}{l} \sin \frac{n\pi x}{l} dx = \frac{l}{2} F_n \tag{7-18.16b}$$

Thus, (7-18.16a) gives

$$\frac{100}{2} A_n = \frac{200}{n\pi}$$

$$A_n = \frac{4}{n\pi} \tag{7-18.16c}$$

Thus, the **general solution** is the sum of the steady-state solution (7-18.9) and the transient solution (7-18.14), as follows:

$$e = e_s + e_T$$

$$= \frac{3x}{50} + \frac{4}{\pi} \sum_{n=1}^{\infty} \left(\frac{1}{n} \sin \frac{n\pi x}{100} \right) e^{-\frac{1}{(CR)} \left(\frac{n\pi}{100} \right)^2 t} \tag{7-18.17}$$

Equation (7-18.2a) gives current as the gradient of (7-18.17) divided by the resistance R as follows

$$i = -\frac{1}{R}\frac{\partial e}{\partial x}$$

$$= -\frac{3}{50R} - \frac{1}{25R}\sum_{n=1}^{\infty}\left(\cos\frac{n\pi x}{100}\right)e^{-\frac{1}{(CR)}\left(\frac{n\pi}{100}\right)^2 t}$$ (7-18.18)

$$= -\frac{1}{2} - \frac{1}{3}\sum_{n=1}^{\infty}\left(\cos\frac{n\pi x}{100}\right)e^{-\frac{n^2\pi^2}{0.0024}t}$$

We note the **decay factor** in the exponential term.

Example 105

Given the case of **Radio Equations** defined by negligible R (resistance) and G (leakage), in the **Telephone equations,** due to **high frequencies**, such that

Radio Equations

$$-\frac{\partial e}{\partial x} = L\frac{\partial i}{\partial t}$$ (7-19.1)

$$\frac{\partial i}{\partial x} = -C\frac{\partial e}{\partial t}$$ (7-19.2)

$$\frac{\partial^2 e}{\partial x^2} = CL\frac{\partial^2 e}{\partial t^2}$$ (7-19.3)

$$\frac{\partial^2 i}{\partial x^2} = CL\frac{\partial^2 i}{\partial t^2}$$ (7-19.4)

Suppose that a line of 200 km long **energized initially** by e.m.f. described as

$$e(x,0) = E\sin\frac{\pi x}{200}$$ (7-19.5)

And **initial constant current** given by

$$i(x,0) = I_0$$ (7-19.6)

Those two conditions only determine the x-derivatives at t=0 (not the t-derivatives) such that

$$\frac{\partial e(x,0)}{\partial x} = \frac{\pi}{200}E\cos\frac{\pi x}{200}$$ (7-19.7)

$$\frac{\partial i(x,0)}{\partial x} = 0$$ (7-19.8)

If the line is grounded at both ends at t = 0, then

$$e(0,t) = e(200,t) = 0 \tag{7-19.9}$$

Thus, we could obtain a solution following the same steps adopted in obtaining equation (7-18.14):

$$e(x,t) = \left(A\sin\frac{\pi x}{200} \right)\cos\frac{\pi t}{200\sqrt{LC}} \tag{7-19.10}$$

We note the **absence of decay factor** and the sinusoidal nature of the voltage in time and space.

Obtaining the **current** from (7-19.10) requires two steps as follows:

First, equation (7-19.1) and (7-19.10) give the rate of change of the current as follows:

$$L\frac{\partial i}{\partial t} = -\frac{\partial}{\partial x}\left(A\sin\frac{\pi x}{200} \right)\cos\frac{\pi t}{200\sqrt{LC}}$$
$$= -\frac{\pi}{200}\left(A\cos\frac{\pi x}{200} \right)\cos\frac{\pi t}{200\sqrt{LC}} \tag{7-19.11a}$$

Second, integrating with respect to t, we get

$$i = -\sqrt{\frac{C}{L}}\left(A\cos\frac{\pi x}{200} \right)\sin\frac{\pi t}{200\sqrt{LC}} + C(x) \tag{7-19.11b}$$

Where, $C(x)$ is an arbitrary function in x, determined from the initial condition (7-19.6), such that

$$i = -\sqrt{\frac{C}{L}}\left(A\cos\frac{\pi x}{200} \right)\sin\frac{\pi t}{200\sqrt{LC}} + I_0 \tag{7-19.11c}$$

Summary of electric wave equations

Telephone Equations

$$\frac{\partial^2 e}{\partial x^2} = CL\frac{\partial^2 e}{\partial t^2} + (GL + CR)\frac{\partial e}{\partial t} + RGe \tag{7-17.3c}$$

309

$$\left(\frac{\partial^2 i}{\partial x^2} - GL\frac{\partial i}{\partial t} - GRi \right) = CL\frac{\partial^2 i}{\partial t^2} + CR\frac{\partial i}{\partial t}$$ (7-17.4d)

Telegraph Equations:

$$e = \frac{3x}{50} + \frac{4}{\pi}\sum_{n=1}^{\infty}\left(\frac{1}{n}\sin\frac{n\pi x}{100} \right)e^{-\frac{1}{(CR)}\left(\frac{n\pi}{100}\right)^2 t}$$ (7-18.17)

$$i = -\frac{1}{2} - \frac{1}{3}\sum_{n=1}^{\infty}\left(\cos\frac{n\pi x}{100} \right)e^{-\frac{n^2\pi^2}{0.0024}t}$$ (7-18.18)

Radio Equations

$$e(x,t) = \left(A\sin\frac{\pi x}{200} \right)\cos\frac{\pi t}{200\sqrt{LC}}$$ (7-19.10)

$$i = -\sqrt{\frac{C}{L}}\left(A\cos\frac{\pi x}{200} \right)\sin\frac{\pi t}{200\sqrt{LC}} + I_0$$ (7-19.11c)

7-2.5. One dimensional heat conduction

Consider the Cartesian geometry of balancing heat flow in a space volume in Figure 7-14.

Let ρ = **density** of conducting medium, c its **specific heat** and K its **thermal conductivity**. Consider an infinitesimal rectangular parallelepiped, whose centre P is the point $P(x,y,z)$ and whose faces are parallel to the coordinate planes and let the lengths of its sides be δx, δy, and δz.

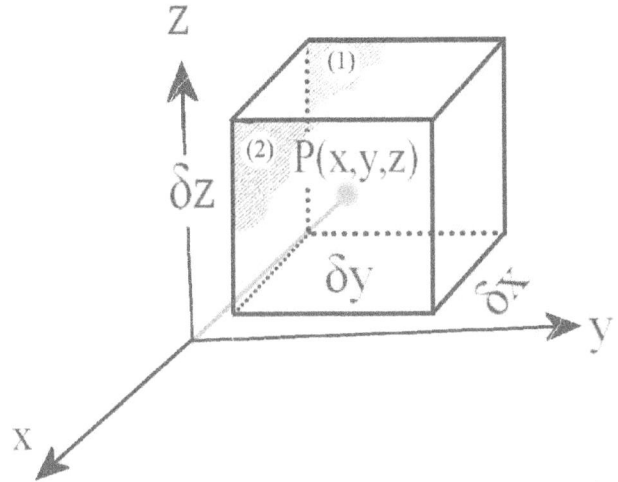

Figure 7-14. Geometry of infinitesimal space volume in Cartesian coordinates for balancing net flow of heat.

The quantity of heat entering the element in time δt through the face (1) is

$$-K\delta x\delta y\left(\frac{\partial U(x,y,z,t)}{\partial x}-\frac{\partial^2 U(x,y,z,t)}{\partial x^2}.\frac{\delta x}{2}\right)\delta t \qquad (7\text{-}20.1)$$

The quantity of heal leaving the element in time δt from face (2) is

$$-K\delta x\delta y\left(\frac{\partial U(x,y,z,t)}{\partial x}+\frac{\partial^2 U(x,y,z,t)}{\partial x^2}.\frac{\delta x}{2}\right)\delta t \qquad (7\text{-}20.2)$$

Hence net gain of heat from the two faces is

$$K\delta x\delta y.\delta t\frac{\partial^2 U(x,y,z,t)}{\partial x^2} \qquad (7\text{-}20.3)$$

We have two similar contributions from the remaining two pairs of parallel faces. Hence total gain of heat in the element is

$$K\delta x\delta y\delta z.\delta t\left(\frac{\partial^2 U(x,y,z,t)}{\partial x^2}+\frac{\partial^2 U(x,y,z,t)}{\partial y^2}+\frac{\partial^2 U(x,y,z,t)}{\partial z^2}\right) \qquad (7\text{-}20.4)$$

This will increase the temperature $U(x, y, z, t)$ by

$$\rho.c.\frac{\partial U(x, y, z, t)}{\partial t}\delta t \qquad\qquad (7\text{-}20.5)$$

Therefore, the balance heat flow versus temperature rise of internal matter in the volume element is

$$K\delta x\delta y\delta z.\delta t\left(\frac{\partial^2 U(x, y, z, t)}{\partial x^2} + \frac{\partial^2 U(x, y, z, t)}{\partial y^2} + \frac{\partial^2 U(x, y, z, t)}{\partial z^2}\right) = \rho.c.\delta x\delta y\delta z.\frac{\partial U(x, y, z, t)}{\partial t}\delta t$$

i.e.,

$$\frac{\partial^2 U(x, y, z, t)}{\partial x^2} + \frac{\partial^2 U(x, y, z, t)}{\partial y^2} + \frac{\partial^2 U(x, y, z, t)}{\partial z^2} = \frac{\rho.c}{K}.\frac{\partial U(x, y, z, t)}{\partial t} \qquad (7\text{-}20.6)$$

$$\nabla^2 U(x, y, z, t) = \frac{\rho.c}{K}.\frac{\partial U(x, y, z, t)}{\partial t} \qquad\qquad (7\text{-}20.7)$$

Put

$$k = \frac{K}{\rho.c} \qquad\qquad (7\text{-}20.8)$$

Where k is known as the **thermal diffusivity** of the material.
∇^2 is **Laplacian operator** defined in the Cartesian coordinates by

$$\nabla^2 \cong \frac{\partial^2}{\partial x^2} + \frac{\partial^2}{\partial y^2} + \frac{\partial^2}{\partial z^2} \qquad\qquad (7\text{-}20.9)$$

Therefore,

$$\frac{\partial U}{\partial t} = k\nabla^2 U \qquad\qquad (7\text{-}20.10)$$

This is known as the **heat equation or equation of diffusion**.

If the flow of heat is **uniflow** i.e., in one direction, say the x - axis then the last equation becomes

$$\frac{\partial U}{\partial t} = k\frac{\partial^2 U}{\partial x^2} \qquad\qquad (7\text{-}20.11)$$

When the flow of heat is **steady** i.e., independent of the time, the heat equation reduces to

$$\nabla^2 U = 0 \qquad\qquad (7\text{-}20.12)$$

312

i.e., the temperature U satisfies **Laplace's equation**.

7-2.6. Heat conduction curvilinear coordinates

Curvilinear Coordinate Variable Parameters (h_1, h_2, h_3)

Curvilinear coordinates are generalization of coordinate systems that could be easily converted to the Cartesian, Cylindrical, Spherical, or otherwise desired system of coordinates.

Consider the Curvilinear geometry of a space volume in Figure 7-15.

Curvilinear generalization is derived as follows.

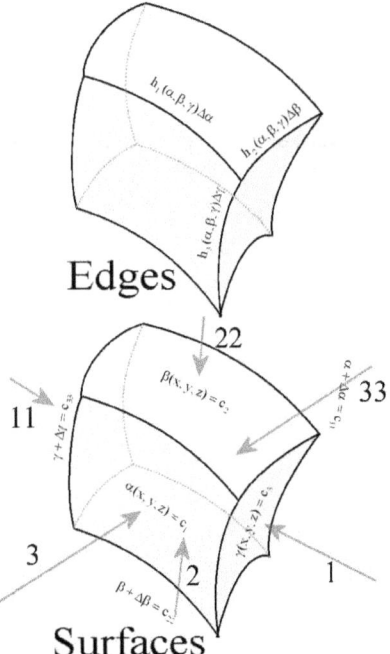

Figure 7-15. Geometry of infinitesimal space volume in Curvilinear coordinates for formulating general transformation between coordinate systems.

Consider the system of three space-surfaces shown in Figure 7-15:

$$\alpha(x, y, z) = c_1$$
$$\beta(x, y, z) = c_2 \tag{7-21.1}$$
$$\gamma(x, y, z) = c_3$$

Consider a system of parallel surfaces to the three above defined by

$$\alpha + \Delta\alpha = c_{11}$$
$$\beta + \Delta\beta = c_{22} \tag{7-21.2}$$
$$\gamma + \Delta\gamma = c_{33}$$

Assume that the displacements between the parallel surfaces c_1-c_{11}, c_2-c_{22}, and c_3=c_{33} are defined by

$$h_1(\alpha, \beta, \gamma)\Delta\alpha$$
$$h_2(\alpha, \beta, \gamma)\Delta\beta \tag{7-21.3}$$
$$h_3(\alpha, \beta, \gamma)\Delta\gamma$$

Where, h_1, h_2, and h_3 are the **curvilinear coordinate variable parameters** corresponding to each coordinate system will be shown to be equal to the following tabulated values:

Systems	Coordinates	h_1	h_2	h_3
Cartesian	x, y, z	1	1	1
Cylindrical	r, θ, z	1	r	1
Spherical	r, θ, φ	1	r sin φ	R

Curvilinear Arc Length

The length of arc is determined by Pythagorean Theorem as

$$(ds)^2 = (h_1\Delta\alpha)^2 + (h_2\Delta\beta)^2 + (h_3\Delta\gamma)^2 \tag{7-21.4}$$

Systems	Length of Arc
Cartesian	$(ds)^2 = (\Delta x)^2 + (\Delta y)^2 + (\Delta z)^2$
Cylindrical	$(ds)^2 = (\Delta r)^2 + (r\Delta\theta)^2 + (\Delta z)^2$
Spherical	$(ds)^2 = (\Delta r)^2 + (r\sin\varphi\Delta\theta)^2 + (r\Delta\varphi)^2$

Curvilinear Laplacian Operator ∇^2

314

Consider the Curvilinear geometry of balancing heat flow in a space volume in Figure 7-16.

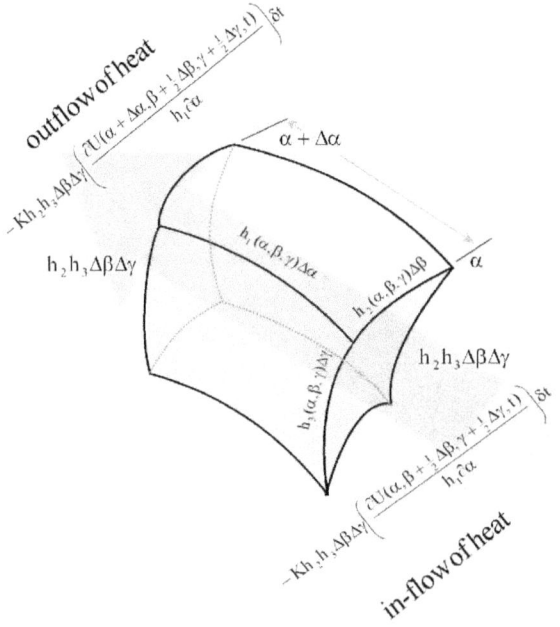

Figure 7-16. Geometry of infinitesimal space volume in Curvilinear coordinates for balancing net flow of heat.

The quantity of heat entering the element in time δt through the faces (c_1) is

$$- Kh_2h_3\,\Delta\beta\,\Delta\gamma\left(\frac{\partial U(\alpha, \beta + \frac{1}{2}\beta, \gamma + \frac{1}{2}\,\Delta\gamma)}{h_1\partial\alpha}\right)\delta t \qquad (7\text{-}22.1)$$

This is the flow of heat from (α) to ($\alpha +\Delta$ α) perpendicular to the area ($h_2h_3\Delta$ β Δ γ) in the middle of each of (β) and ($\beta+\Delta$ β) and (γ) and ($\gamma +\Delta$ γ) distances.

Keep in mind that h_1, h_2, and h_3 are variable parameters dependent to α, β, and γ.

315

The quantity of heal leaving the element in time δt from face (c_{11}) is

$$- K \Delta\beta \, \Delta\gamma \left(\frac{h_2 h_3 \partial U}{h_1 \partial \alpha} \right)_{\cdot (\alpha + \Delta\alpha, \beta + \frac{1}{2}\beta, \gamma + \frac{1}{2}\Delta\gamma)} \delta t \qquad (7\text{-}22.2)$$

Hence, net gain of heat from the two faces is

$$K \, \Delta\alpha \, \Delta\beta \, \Delta\gamma \left(\frac{\left(\dfrac{h_2 h_3}{h_1} \dfrac{\partial U}{\partial \alpha} \right)_{(\alpha + \Delta\alpha, \beta + \frac{1}{2}\beta, \gamma + \frac{1}{2}\Delta\gamma)} - \left(\dfrac{h_2 h_3}{h_1} \dfrac{\partial U}{\partial \alpha} \right)_{(\alpha, \beta + \frac{1}{2}\beta, \gamma + \frac{1}{2}\Delta\gamma)}}{\Delta\alpha} \right) . \delta t \qquad (7\text{-}22.3)$$

Note that we multiplied and divided by the same quantity $(\Delta\,\alpha)$ in order to obtain an expression in the element of volume. Not also that we have dismissed second-order derivatives of U.

We have two similar contributions from the remaining two pairs of parallel faces. Hence total gain of heat in the element is

$$K \, \Delta\alpha \, \Delta\beta \, \Delta\gamma . \delta t \left(\frac{\partial}{\partial \alpha} \left(\frac{h_2 h_3}{h_1} \frac{\partial U}{\partial \alpha} \right) + \frac{\partial}{\partial \beta} \left(\frac{h_1 h_3}{h_2} \frac{\partial U}{\partial \beta} \right) + \frac{\partial}{\partial \gamma} \left(\frac{h_2 h_1}{h_3} \frac{\partial U}{\partial \gamma} \right) \right) \qquad (7\text{-}22.4)$$

This will increase the temperature $U(x,y,z,t)$ by

$$\rho . c_p . \Delta\alpha \, \Delta\beta \, \Delta\gamma . \delta t \frac{\partial U(\alpha, \beta, \gamma, t)}{\partial t} \qquad (7\text{-}22.5)$$

Where, ρ is the density of the medium, c_p its constant **specific heat**.

Therefore, the **balance heat flow** versus temperature rise of internal matter, in the volume element, is

$$K \, \Delta\alpha \, \Delta\beta \, \Delta\gamma . \delta t \left(\frac{\partial}{\partial \alpha} \left(\frac{h_2 h_3}{h_1} \frac{\partial U}{\partial \alpha} \right) + \frac{\partial}{\partial \beta} \left(\frac{h_1 h_3}{h_2} \frac{\partial U}{\partial \beta} \right) + \frac{\partial}{\partial \gamma} \left(\frac{h_2 h_1}{h_3} \frac{\partial U}{\partial \gamma} \right) \right)$$

$$= \rho . c_p . \Delta\alpha \, \Delta\beta \, \Delta\gamma . \delta t \frac{\partial U(\alpha, \beta, \gamma, t)}{\partial t}$$

i.e.,

316

$$\left(\frac{\partial}{\partial\alpha} \left(\frac{h_2 h_3}{h_1} \frac{\partial}{\partial\alpha} \right) + \frac{\partial}{\partial\beta} \left(\frac{h_1 h_3}{h_2} \frac{\partial}{\partial\beta} \right) + \frac{\partial}{\partial\gamma} \left(\frac{h_2 h_1}{h_3} \frac{\partial}{\partial\gamma} \right) \right) U(\alpha,\beta,\gamma,t) = \frac{\rho.c_p}{K} \frac{\partial U(\alpha,\beta,\gamma,t)}{\partial t} \qquad (7\text{-}22.6)$$

Put in a common symbolic notation, the **curvilinear heat equation** is

$$\nabla^2 U(\alpha,\beta,\gamma,t) = \frac{1}{k} \cdot \overset{..}{U}(\alpha,\beta,\gamma,t) \qquad (7\text{-}22.7)$$

Where k is the **thermal diffusivity** of the material defined by

$$k = \frac{K}{\rho.c} \qquad (7\text{-}22.8)$$

∇^2 is **Laplacian operator** defined in the Curvilinear coordinates by

$$\nabla^2 \cong \frac{\partial}{\partial\alpha} \left(\frac{h_2 h_3}{h_1} \frac{\partial}{\partial\alpha} \right) + \frac{\partial}{\partial\beta} \left(\frac{h_1 h_3}{h_2} \frac{\partial}{\partial\beta} \right) + \frac{\partial}{\partial\gamma} \left(\frac{h_2 h_1}{h_3} \frac{\partial}{\partial\gamma} \right) \qquad (7\text{-}22.9)$$

7-2.7. Steady-state heat conduction

Example 106

Consider a heat conducting plate with rectangular cross-section in the xy-plane exposed to the following boundary conditions.

$U(0,y) = U(20,y) = 0$	for all values of $y \geq 0$	(7-23.1a)
$U(x,\infty) = 0$		(7-23.1b)
$U(x,0) = 50$	for all values of $20 \geq x \geq 0$	(7-23.1c)

Find the temperature distribution $U(x,y)$.

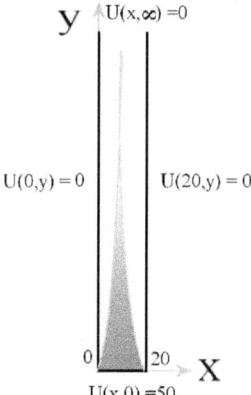

Figure 17. Geometry of steady-state two-dimensional heat flow.

Solution

The described problem is two dimensional and time-independent, that can be solved by equation (7-20.12) in the xy-plane as follows:

$$\frac{\partial^2 U}{\partial x^2} + \frac{\partial^2 U}{\partial y^2} = 0 \qquad (7\text{-}23.2)$$

We have solved this equation by separation of variables in equation (6-6.7) and discussed few available solutions of the form:

$$U(x,y) = \left(C_1 e^{kx} + C_2 e^{-kx}\right)\left[C_3 \cos ky + C_4 \sin ky\right] \qquad (7\text{-}23.3a)$$
$$U(x,y) = \left(C_5 e^{ky} + C_6 e^{-ky}\right)\left[C_7 \cos kx + C_8 \sin kx\right] \qquad (7\text{-}23.3b)$$
$$U(x,y) = \left(C_9 y + C_{10}\right)\left[C_{11} x + C_{12}\right] \qquad (7\text{-}23.3c)$$

Since all three solutions satisfy the heat equation (7-23.2), we settle on those that satisfy all boundary conditions (7-23.1).

Apparently, equation (7-23.3b) is the only solution that decays to zero at infinite y, required by equation (7-23.1b). Therefore, we will determine the four constants in (7-23.3b) by substituting by the boundary conditions as follows:

$$U(0,y) = \left(C_5 e^{ky} + C_6 e^{-ky}\right)[C_7] = 0 - - - - - - - - - - - > C_7 = 0$$

$$U(20,y) = \left(C_5 e^{ky} + C_6 e^{-ky}\right)[C_8 \sin 20k] = 0 - - - - - - > 20k = n\pi, \quad n = 1,2,3...$$

$$U(x,\infty) = \left(C_5 e^{ky}\right)[C_7 \cos kx + C_8 \sin kx] - - - - - - - - > C_5 = 0 \qquad (7\text{-}23.4)$$

$$U(x,0) = \left(C_6\right)\left[C_8 \sin \frac{n\pi}{20} x\right] = 50$$

This time, we executed the boundary conditions expediently, eliminated C_7 as the coefficient of cosine that must vanish, eliminated C_5 as the coefficient of a growth exponential term that is rejected by boundary conditions, equated (20k) by the roots of the sine function and left the constants $C_6 C_8$ to be determined by the common **Euler's formula** discussed in equation (6-7.10d), as follows

$$\int_{x=0}^{l} F_n \sin \frac{m\pi x}{l} \sin \frac{n\pi x}{l} dx = \frac{l}{2} F_n \qquad (7\text{-}23.5)$$

Thus, the last equation in (7-23.4) is multiplied by $\sin \dfrac{n\pi x}{20}$, on both sides, and integrated to give

$$\int_0^{20} A_n \sin \frac{n\pi}{20} x \sin \frac{m\pi}{20} x dx = 50 \int_0^{20} \sin \frac{m\pi}{20} x dx \qquad (7\text{-}23.6)$$

$$10 A_n = 50 \frac{20}{m\pi}\left[-\cos \frac{m\pi}{20} x\right]_0^{20}$$

$$= 1000 \frac{1}{m\pi}[+1+1] - - - m - is - odd \qquad (7\text{-}23.7a)$$

$$= 1000 \frac{1}{m\pi}[-1+1] - - - m - is - even$$

$$A_n = 200 \frac{1}{m\pi}, \text{ for odd m} \qquad (7\text{-}23.7b)$$

The general solution is obtained by substituting the constants in equation (7-23.3b) as follows

$$U(x,y) = \sum_{n=1,3,4,}^{\infty} \frac{200}{n\pi} e^{-\frac{n\pi}{20} y} \sin \frac{n\pi}{20} x \qquad (7\text{-}23.8)$$

We should note that the two boundary conditions $U(x,0) = 50$ and $U(0,y) = 0$ cannot be achieved in any practical situation because when both x and y are zeros, we have the contradiction of two different temperatures in the same spot $(x,y) = (0,0)$.

319

7-2.8. One dimensional variable heat conduction

Example 107

Given a thin, 10 cm long heat conductor, Figure 7-18, that is kept insulated on all sides. Initially, the rod has temperature profile given by

$$U(x,0) = 2x \qquad (7-24.1)$$

Where x is measures from one end of the rod. The rod is then exposed to 0°C temperatures at both ends.

Assume that the thermal diffusivity k = 5/3, describe the temporal variation of temperature versus x.

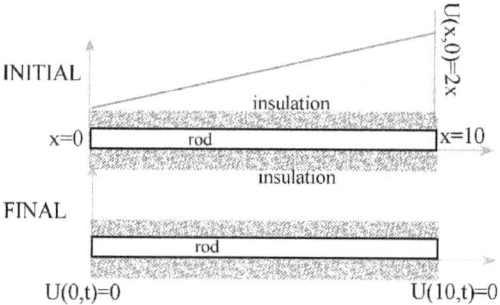

Figure 7-18. Geometry of variable one-dimensional heat flow.

Solution

Equation (7-20.11) describes one-dimensional time-dependent heat flow as follows

$$\frac{\partial U}{\partial t} = k \frac{\partial^2 U}{\partial x^2} \qquad (7-24.2)$$

The boundary and initial conditions are

$$U(x,0) = 2x \qquad (7-24.3a)$$
$$U(0,t) = U(10,t) = 0 \qquad (7-24.3b)$$
$$l = 10 \qquad (7-24.3c)$$
$$k = 5/3 \qquad (7-24.3d)$$

320

Using the separation of variables, equation (6-7.7) gives the following particular solutions

$$u(x,t) = X(x)T(t)$$
$$= e^{-\lambda kt} \left[C \cos \sqrt{\lambda} x + F \sin \sqrt{\lambda} x \right]$$

(7-24.4)

Substituting by the boundary and initial conditions, we get

$$u(x,0) = \left[C \cos \sqrt{\lambda} x + F \sin \sqrt{\lambda} x \right] = 2x$$
$$u(0,t) = e^{-\lambda kt} \left[C \right] = 0 ------------- > C = 0$$
$$u(10,t) = e^{-\lambda kt} \left[F \sin \sqrt{\lambda} 10 \right] = 0 ------- > \sqrt{\lambda} 10 = n\pi$$

(7-24.5)

So far, we have determined two out of three constants as follows

$$u(x,t) = \sum_{n=1}^{\infty} F_n e^{-\frac{n^2\pi^2}{100}kt} \sin \frac{n\pi}{10} x$$

(7-24.6)

Thus, the remaining coefficient F is determined by the same Euler's method, equation (7-23.5), by multiplying both sides of the first equation in (7-24.5) by $\sin \frac{m\pi}{10}$ and integrating both sides from $x = 0$ to $x = 10$, as follows

$$\sum_{n=1}^{\infty} \int_0^{10} F_n \sin \frac{m\pi}{10} x \sin \frac{n\pi}{10} x\, dx = 2 \int_0^{10} x \sin \frac{m\pi}{10} x\, dx$$

(7-24.7)

Performing the integration, we get

$$5F_n = -\frac{20}{m\pi} \left(\int_0^{10} x d \cos \frac{m\pi}{10} x \right)$$
$$= -\frac{20}{m\pi} \left[x \cos \frac{m\pi}{10} x - \frac{10}{m\pi} \sin \frac{m\pi}{10} x \right]_0^{10}$$

(7-24.8a)

$$= \frac{200}{m\pi} ------- > m - odd$$
$$= -\frac{200}{m\pi} ------ > m - even$$

$$F_n = (-1)^{n+1} \frac{40}{n\pi}$$

(7-24.8b)

Noting that m must be equal to n in order for the integration to exist (has value other than zero).

Thus, the final solution is

$$u(x,t) = \frac{40}{\pi} \sum_{n=1}^{\infty} F_n = (-1)^{n+1} \frac{1}{n} e^{-\frac{n^2 \pi^2}{60} t} \sin \frac{n\pi}{10} x \qquad (7\text{-}24.6)$$

Where k is substituted by its value 5/3.

Example 108

Given a thin, 4 cm long heat conductor that is kept insulated on all sides, Figure 7-19. Initially, one end of the rod was kept at 20 °C, the other at 60 °C.

The rod is then exposed to sudden change where the 20°C was lowered to 10 °C and 60 °C raised to 70 °C.

Assume that the thermal diffusivity k = 0.2, describe the temporal variation of temperature versus x.

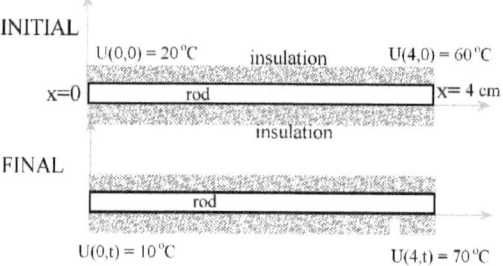

Figure 7-19. Geometry of variable one-dimensional heat flow where ends are kept over zero degrees.

Solution

Equation (7-20.11) describes one-dimensional time-dependent heat flow as follows

$$\frac{\partial U}{\partial t} = k \frac{\partial^2 U}{\partial x^2} \qquad (7\text{-}25.1)$$

The **boundary and initial conditions** are

$$U(0,0) = 20\,^{\circ}C \tag{7-25.2a}$$
$$U(4,0) = 60\,^{\circ}C \tag{7-25.2b}$$
$$U(0,t) = 10\,^{\circ}C \tag{7-25.2c}$$
$$U(4,t) = 70\,^{\circ}C \tag{7-25.2d}$$

$$l = 4\ cm \tag{7-25.2e}$$
$$k = 0.2 \tag{7-25.2f}$$

Steady-States

The initial steady-state implies constant temperature gradient between the two extreme ends of the rod such that

$$\frac{U(x,0) - 20}{x - 0} = \frac{60 - 20}{4 - 0} \tag{7-25.3a}$$

Or

$$U(x,0) = 10x + 20 \tag{7-25.3b}$$

Similarly, the final steady-state implies a different constant temperature gradient between the two extreme ends of the rod such that

$$\frac{U_S(x,t) - 10}{x - 0} = \frac{70 - 10}{4 - 0} \tag{7-25.4a}$$

Or

$$U_S(x,t) = 15x + 10 \tag{7-25.4b}$$

This is time-independent **final** steady-state solution.

Homogenization of the boundary conditions

The **transient solution** U_T is now subjected to **boundary conditions** (7-25.2c), (7-25.2d), and

(7-25.4b)

$$U_T(0,t) = U(0,t) - (15x + 10)$$
$$= 10 - (15x0 + 10) = 0$$
$$U_T(4,t) = U(4,t) - (15x + 10) \tag{7-25.4c}$$
$$= 70 - (15x4 + 10) = 0$$

The transient solution U_T is subjected to **initial conditions** equation (7-25.3b) and (7-25.4b)

$$U_T(x,0) = U(x,0) - (15x + 10)$$

$$= (10x + 20) - (15x + 10) \qquad (7\text{-}25.4d)$$

$$= -5x + 10$$

Now, before solving the p.d.e. of heat flow, (7-25.1), we have devised a transient solution that homogenizes the p.d.e. that are easy to solve. The transient solution achieves transition between changing the temperatures at the ends of the rod to reaching steady-state final temperature profile. The transient solution also satisfies the transition between the initial steady state of U=10x+20 to the final steady-state of U =15x+10. Figure 7-20.

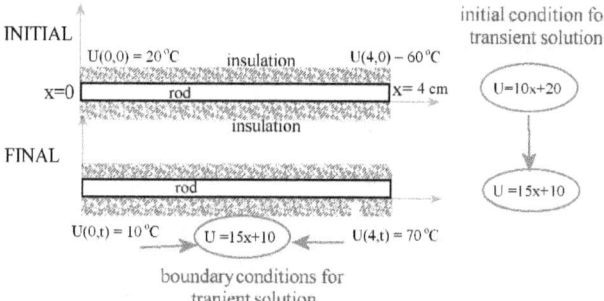

Figure 7-20. Initial and boundary conditions for transient temperature distribution required to solve the p.d.e. of heat flow.

Solution of the p.d.e. of heat flow

Using the separation of variables, equation (6-7.7) gives the following particular solutions

$$u(x,t) = X(x)T(t)$$

$$= e^{-\lambda k t}\left[C\cos\sqrt{\lambda}x + F\sin\sqrt{\lambda}x\right] \qquad (7\text{-}25.5)$$

Substituting by the boundary and initial conditions, we get

$$u(0,t) = e^{-\lambda k t}\left[C\right] = 0 ------------ > C = 0$$

$$u(4,t) = e^{-\lambda k t}\left[F\sin\sqrt{\lambda}4\right] = 0 --------- > \sqrt{\lambda}4 = n\pi \qquad (7\text{-}25.6)$$

$$u(x,0) = \left[F\sin\frac{n\pi}{4}x\right] = -5x + 10$$

We could see from the first two equation of (7-25.6) that homogenizing the boundary conditions zeroed the RHS and eliminated any additional constants.

Thus, the remaining coefficient F is determined by the same Euler's method, equation (7-23.5), by multiplying both sides of the third equation in (7-25.6) by $\sin\frac{m\pi}{4}$ and integrating both sides from $x = 0$ to $x = 4$, as follows

$$\sum_{n=1}^{\infty} \int_0^4 F_n \sin\frac{m\pi}{4} x \sin\frac{n\pi}{4} x dx = \int_0^4 (-5x+10)\sin\frac{m\pi}{4} x dx \qquad (7-25.7)$$

Performing the integration, we get

$$2F_n = -5\int_0^4 x \sin\frac{m\pi}{4} x dx + 10\int_0^4 \sin\frac{m\pi}{4} x dx$$

$$= -\frac{20}{m\pi}\left[\left[-x\cos\frac{m\pi}{4} x - \frac{4}{m\pi}\sin\frac{m\pi}{4} x + 2\cos\frac{m\pi}{4} x\right]_0^4\right] \qquad (7-25.8a)$$

$$= 0 - - - - - - m - is - odd$$

$$= \frac{80}{m\pi} - - - - - m - is - even$$

$$F_n = \frac{40}{m\pi} - - - m - even \qquad (7-25.8b)$$

Since m is even, we put n = 2m in the final transient solution such that

$$u_T(x,t) = \frac{40}{\pi}\sum_{n=1}^{\infty}\frac{1}{2n}e^{-\frac{4n^2\pi^2}{80}t}\sin\frac{2n\pi}{4} x$$

$$= \frac{20}{\pi}\sum_{n=1}^{\infty}\frac{1}{n}e^{-\frac{n^2\pi^2}{20}t}\sin\frac{n\pi}{2} x \qquad (7-25.9)$$

Where k is substituted by its value 0.2.

The final general solution is comprised of the sum of the transient solution u_T (7-25.9) and final steady-state solution (7-25.4b) as follows:

$$u(x,t) = u_S(x,t) + u_T(x,t)$$

$$= (15x+10) + \frac{20}{\pi}\sum_{n=1}^{\infty}\frac{1}{n}e^{-\frac{n^2\pi^2}{20}t}\sin\frac{n\pi}{2} x \qquad (7-25.10)$$

7-3. Exercise on Linear Partial Differential Equations

7-1.   Solve the partial differential equation

$$U_{xx}(x,t) + U_{tx}(x,t) - 2\, U_{tt}(x,t) = 0 \qquad\qquad (x > 0\,, t > 0\,)$$

subject to the following boundary conditions

        $U(x,0) = U_t(x,0) = 0$
        $U(x,t) = 0$ as $x \to \infty$
        $U(0,t) = F(t)$

7-2.   Solve the following boundary value problem

$$Y_x(x,t) + x\, Y_t(x,t) = 0$$

subject to the following boundary conditions

        $Y(x,0) = 0$
        $Y(0,t) = t$

7-3.   Solve the partial differential equation

$$Y_{tt}(x,t) = a^2\, Y_{xx}(x,t) \qquad\qquad\qquad x > 0,\, t > 0$$

subject to the following boundary conditions

        $Y(x,0) = 0$
        $Y_t(x,0) = -u_o$
        $Y(0,t) = 0$
        $Y_x(x,t) = 0$ as $x \to \infty$

7-4.   Find a bounded solution of

$$x\frac{\partial U}{\partial x} + \frac{\partial U}{\partial y} = xe^{-y} \qquad\qquad 0 < x < 1,\, y > 0$$

Which satisfies $U(x,0) = x$, $0 < x < 1$

7-5.   Solve the equation

$$\frac{\partial U}{\partial t} + x\frac{\partial U}{\partial x} + U = x \qquad\qquad x > 0,\, t > 0$$

with the boundary conditions

$U(x,0) = 0$
$U(0,t) = 0$

7-6.   Show that *in* solving the partial differential equation

$$\frac{\partial^2 Y}{\partial t^2} = a^2 \frac{\partial^2 Y}{\partial x^2}$$

We can solve the equation

$$\frac{\partial^2 Y}{\partial t^2} = \frac{\partial^2 Y}{\partial x^2}$$

then in the result, we replace t by at.

7-7.   Solve the equation

$$\frac{\partial^2 Y}{\partial x^2} = 16 \frac{\partial^2 Y}{\partial t^2} \qquad\qquad x > 0, t > 0$$

subject to the following boundary conditions

$Y(x,0) = 0$
$Y_t(x,0) = -1$
$Y(0,t) = t^2$
$Y_x(x,t)$ exists as $x \to \infty$ for fixed $t > 0$.

7-8.   Solve the equation

$$\frac{\partial Y}{\partial x} + 4\frac{\partial Y}{\partial t} = -8t \qquad\qquad x > 0, t > 0$$

subject to the following boundary conditions

$Y(x,0) = 0$
$Y(0,t) = 2t^2$

7-9.   Solve the equation

$$\frac{\partial Y}{\partial x} + 2\frac{\partial Y}{\partial t} = 4t \qquad\qquad x > 0, t > 0$$

subject to the boundary conditions

$Y(x,0) = 0$
$Y(0,t) = 2t^3$

7-10.   Solve the equation

$$\frac{\partial^2 Y}{\partial t^2} = 4\frac{\partial^2 Y}{\partial x^2} \qquad\qquad x > 0, t > 0$$

subject to the boundary conditions

$Y(x,0) = 0$
$Y_t(x,0) = 2$
$Y(0,t) = \sin t$
$Y_x(x,t)$ exists as $x \to \infty$ for fixed $t > 0$.

Answers

7-1. $U(x,t) = 0$ $0 < t < 2x$
       $U(x,t) = F(t - 2x)$ $t > 2x$]

7-2. $Y(x,t) \quad = 0$ $0 < t < x^2/2$
               $= t - x^2/2$ $t > x^2/2$]

7-3. $Y(x,t) \quad = -u_0$ $0 < t < x/a$
               $= -u_0 x/a$ $t > x/a$]

7-4. $U(x,y) = xe^{-y}(1+y)$

7-5. $U(x,t) = U(x,t) = \dfrac{1}{2}x\left(1 - e^{-2t}\right)$

7-6.   Proof required

7-7. $Y(x,t) \quad = -t$ $0 \le t \le 4x$
               $= (t - 4x)^2 - 4x$ $t \ge 4x$

7-8. $Y(x,t) \quad = -t^2$ $0 < t < 4x$
               $= -t^2 + 3(t - 4x)^2$ $t > 4x$

7-9. $Y(x,t) \quad = t^2$ $0 \le t \le 2x$
               $= t^2 + 2(t - 2x)^3 - (t - 2x)^2$ $t \ge 2x$

7-10. $Y(x,t) \quad = 2t$ $0 \le t \le x/2$
               $= 2t + \sin(t - x/2) - 2(t - x/2)$ $t \ge x/2$

328

Index

329

=================== End Of Book =========================

www.ingramcontent.com/pod-product-compliance
Lightning Source LLC
Chambersburg PA
CBHW051209170526
45166CB00005B/1824